Spectral Methods in Econometrics

SPECTRAL
METHODS
IN
ECONOMETRICS

George S. Fishman

Harvard University Press
Cambridge, Massachusetts   1969

© Copyright 1968, 1969 by The RAND Corporation
All rights reserved
Distributed in Great Britain by Oxford
  University Press, London
Library of Congress Catalog Card
  Number: 72-78517
SBN 674-83191-8
Printed in the United States of America

Preface

Time series analysis is a branch of statistics whose scope and method have broadened considerably in the past quarter century. Among the developments of this period, the spectral approach, which concerns the decomposition of a time series into frequency components, has become a principal tool of analysis in such diverse fields as communications engineering, geophysics, oceanography, and electroencephalography. Spectral methods have also been applied to the analysis of economic time series, and it is evident that the recent extension of these methods to the estimation and testing of distributed lag models will attract more interest in applying this approach to econometrics.

This book describes spectral methods and their use in econometrics. It is intended as an introduction for graduate students and econometricians who wish to familiarize themselves with the topic.

The literature often emphasizes the ability of the spectral approach to reveal periodic or almost periodic components in a time series. While this information is of interest to the econometrician, it does not justify his learning more than a meager amount about the spectral approach to time series analysis, but a number of other reasons do justify learning about it in detail. The spectral approach has conceptual advantages for conveying information about interdependent events and also has desirable properties for working with sample data. In addition, it often suggests models for explaining the time-varying behavior of economic phenomena, it often enables one to determine the effects of transformations on variables, it often simplifies hypothesis testing, and

v

it often permits more efficient use of sample data in the estimation of time-domain model parameters than do more traditional methods.

Mathematically, time series may be regarded as a topic in Hilbert space analysis. If we were to use this approach, the conciseness of the notation would unfortunately obscure many intuitive analogies that prove so helpful in acquainting readers with a new topic. Alternatively, the benefits of spectral methods would not emerge clearly in a predominantly heuristic description. This is especially true in studying simultaneous distributed lag equations. To convey the significance of spectral methods for econometric analyses properly, some compromise is necessary.

A study of this book therefore requires a knowledge of probability theory and difference equations for univariate and bivariate time series analysis, and also a knowledge of matrix analysis and multivariate statistical analysis for multivariate time series analysis. A familiarity with operations on complex variables is also helpful, but not essential.

In writing this book, I have attempted to include as many of the recent advances in spectral methods as have appeared in the open literature and that seem relevant for econometrics. The rate at which new work appears has made this attempt difficult, as have the many new questions that arise whenever spectral methods are extended. For example, the sample properties of the fast Fourier transform technique remain to be worked out in detail. The theory of testing hypotheses related to distributed lag models is also incomplete. Wherever these new advances are mentioned, I have emphasized the transitional state of their development to encourage the reader to watch for future developments.

Thanks are due Marc Nerlove, who read the entire manuscript and suggested numerous improvements, to Bennett Fox, who checked a number of the mathematical results, and to Murray A. Geisler and Albert Madansky, who encouraged me to write the book.

I would also like to thank Jacque McMullen, Nancy Gibson, and Marie Spaugy for their excellent typing of several versions of the manuscript, Doris Dong for preparing the artwork, and Roberta Schneider and Willard Harriss for their editorial assistance.

I am indebted to The RAND Corporation for support of this work under its broad research contract with the United States Air Force.

George S. Fishman

Santa Monica, California
January 1969

Contents

Tables

Figures

1 Introduction

This study describes the use of spectral methods in the analysis of economic time series. These methods form an integral part of modern time series analysis and have been applied to problems in communications engineering, electroencephalography, oceanography, seismography and, more recently, econometrics. The spectral or frequency-domain approach has a twofold value. Firstly, it lends a conceptual simplicity to the theoretical interpretation of time-varying behavior. This promotes an understanding of the implied behavior of time-domain models, especially when linear transformations are involved (Chap. 2). Secondly, the spectral approach has simpler sampling properties than the more straightforward time-domain approach does. This permits a more incisive quantitative description of intertemporal dependence to be drawn from a given record of sample data in the frequency domain than in the time domain (Chap. 3). Intertemporal dependence refers to the associations among the past, present, and future of a time-varying phenomenon.

The methods described here relate to the class of covariance stationary stochastic processes.† A stochastic process $\{X(t), -\infty \leq t \leq \infty\}$ is said to be covariance stationary if it has a time invariant mean, finite mean-square variation, and the product-moment lagged correlation between $X(t)$ and $X(s)$ is a function only of the separation $(t - s)$

---

† A brief description of spectral methods applicable to other classes of stochastic processes may be found in Ref. 7, pp. 350–369.

for all times t and s. When trend is present in a time series, a judiciously chosen linear transformation can improve the approximation to covariance stationarity, permitting spectral methods to be applied.

While the covariance stationarity restrictions are often implicitly imposed on econometric analyses, the value of the spectral approach has generally remained unexploited. This book is intended to describe a number of spectral results for covariance stationary processes and to show how they relate to econometric problems.

Econometrics and modern time series analysis share the common purpose of predicting time-varying behavior, but each emphasizes different considerations. In econometrics the construction of prediction models receives considerable attention [45]. Economic theory, technological laws of transformation, institutional rules and identities all contribute to model specification. Economic theory is the principal contributor and to a large extent dictates the directions of causality among economic phenomena. To estimate the parameters of the resulting models, econometricians have developed a number of methods that are properly regarded as a branch of multivariate statistical analysis.

Modern time series analysis, which is also a branch of multivariate statistical analysis, concentrates on three problem areas to accomplish prediction. The first problem is to develop statistical population measures that describe the central tendencies of time-varying phenomena. The second is to formulate mathematical prediction models that explicitly recognize associations among these phenomena as well as the nature of their central tendencies. The third is to develop estimation and testing procedures for the descriptive population measures and the parameters of the prediction models.

Since modern time series analysis is principally statistical, it concentrates on generalities of associative behavior, of which causal behavior dictated by economic theory in econometrics is but a part. While it is common practice to regard time series analysis as concerning description and econometrics as concerning explanation, the distinction is not clear-cut. We prefer to regard the difference as one of emphasis rather than content.

The relatively weak restrictions of covariance stationarity, when combined with other properties such as ergodicity, permit formal solution of many time series problems. We briefly describe one unusually important property that distinguishes the analysis of covariance stationary time series from other multivariate analyses. Many time-varying phenomena exhibit intertemporal dependence so that, in addition to associations with other phenomena, their future behavior is also related to their past and present behavior. Intertemporal dependence is not itself a unique characteristic, for we may regard a particular phenomenon at different points in time as a set of variables, each associated with a particular point in time. As such, the problem fits into the general multivariate framework.

By noting certain common tendencies among these variables, we may introduce certain simplifications into the model-building, estimating, and testing procedures that reduce the dimensionality of the problem. For example, if we are considering the relationships among m phenomena for n time periods, we are in effect studying mn variables. Taking advantage of the common features of the phenomenon over the n time periods, we need concern ourselves only with relationships among the m phenomena. Covariance stationary processes enable us to take advantage of these dimension reducing techniques. The population measures used are the autocovariance and covariance functions in the time domain, and their respective equivalents (in information) in the frequency domain—the spectrum and cross spectrum.

1.1. Modern Time Series Analysis

Historically, the time-domain approach to time series analysis has been an extension of the classical theory of correlation, the linear association between $X(t)$ and $X(s)$ being measured by the product-moment lagged correlation. The most notable early contributors were Slutzky [88] and Yule [108], who raised a number of interesting questions about the nature of intertemporal dependence that the existing theory could not adequately answer. For example, how was it possible, as Slutzky empirically demonstrated, to induce regular oscillations in a time series through a superposition of purely random events?

The formal theory of modern time series analysis dates from the work of Khintchine [54] and Wiener [102]. By extending the classical Fourier harmonic analysis, they provided a fundamental understanding of the relationship between the autocorrelation function (the product-moment correlation for all values of $t - s$) of a stochastic process and its Fourier transform, the spectral density function. These functions were first defined for a strictly stationary process, which requires that the probability law governing the process be invariant with respect to historical time. Later, the results were extended to the wider class of covariance stationary processes.

Using the framework of stationary processes, Kolmogorov [56] and Wiener [103] produced a formal prediction theory that is the basis of most present work on prediction problems. Whittle [101] has concisely described these prediction techniques and, in particular, how they may be applied to a number of economic and control problems.

The spectrum (the unnormalized spectral density) is the principal function of interest in frequency-domain analysis. It is essentially a harmonic decomposition of variance, and its significance for scientific inquiry had been apparent long before Khintchine's and Wiener's work. In 1898 Schuster [85], searching for hidden periodicities in sunspot data, developed the method of *periodogram analysis* to estimate the spectrum. In economics periodogram analysis was used by Moore [64] in 1914 to study rainfall in the Ohio Valley and by Beveridge [8] in 1922 to study wheat prices in Western Europe. Unfortunately, the method was time consuming and lacked certain desirable statistical properties. But more importantly the underlying probability model was not well understood, and therefore the method led to incorrect conclusions.

The development of the theory of covariance stationary processes broadened the scope of problems to be treated from those containing only regularly periodic components to those exhibiting irregular cyclic behavior. The growth of communications engineering and related disciplines moreover brought a renewed emphasis to the study of the spectrum.

To improve the sampling properties in spectrum estimation, Bartlett [5] and Tukey [11], among others, suggested an approach that emphasized the use of the weighted sample autocovariance function. The improved sampling properties were, in fact, obtained using a smaller number of sample autocovariances than was used in periodogram analysis, a result that significantly reduced computing time. The approach is called *spectrum analysis*, to distinguish it from periodogram analysis, and is presently the commonly employed method of estimating the spectrum.

From its development, spectrum analysis was known to be statistically equivalent to averaging over the periodogram (Section 3.9), but the time-consuming periodogram computation, even on high-speed digital computers, discouraged averaging in the frequency domain in favor of weighting in the time domain. In 1965 Cooley and Tukey [13] described an algorithm, the *fast Fourier transform* technique, that significantly reduced the computing time for periodogram ordinates. As a consequence, periodogram averaging can now save considerable computing time over the *statistically equivalent* weighting of the sample autocovariance function. The saving, which is especially dramatic for long time series, promises to make a *modified periodogram analysis* approach to spectrum analysis more common in the future. Since the literature on the new approach is meager, this study emphasizes the weighted sample autocovariance method. Section 3.10, however, describes a number of major considerations in a modified periodogram analysis approach.

Bartlett's and Tukey's work also revealed that spectrum analysis has considerably simpler sampling properties than does the corresponding time-domain analysis. Hence, given that the autocorrelation and spectral density functions convey the same information, the choice of sample function to use for descriptive and manipulative purposes becomes a matter of statistical convenience.

The refinement of spectrum analysis enabled investigators to acquire a detailed knowledge of the harmonic content of their sample data. The new knowledge was used to improve methods of recording data, to develop models that adequately accounted for the observed harmonic content, and to assist in the removal of unwanted harmonic components.

At first, econometric studies using spectrum analysis exploited its ability to reveal the degree of linear association between economic phenomena and the effects of transformations or variables. Hannan [37, 39, 40] and Nerlove [67], for example, used spectrum analysis to measure the effects of a seasonal adjustment procedure on economic time series. Since then the extension of spectral methods to the estimation and testing of sets of distributed lag equations has significantly broadened the range of amenable econometric problems. The newly developed techniques, which are principally due to Hannan [38, 42], enable one to derive asymptotically efficient coefficient estimators with relatively simple sampling properties. Wallis [96, 97] has used these techniques to study inventory problems.

1.2. Plan of the Study

Each of the four remaining chapters of this study covers a particular topic in time series analysis. Chapter 2 introduces the theory of covariance stationary stochastic processes, placing special emphasis on its application in econometrics. Univariate, bivariate, and multivariate analysis are presented in that order—the last two topics being discussed in terms of distributed lag models.

Chapter 3 describes estimation procedures and approximating sampling distributions applicable to the spectrum, the cross spectrum, and the bivariate frequency response function. Multivariate estimation is next described and several likelihood ratio tests are discussed. As a basis for comparison, the chapter initially describes the sampling properties of the autocovariance and autocorrelation functions.

Chapter 4 describes the spectral approach to estimation and testing for a system of simultaneous distributed lag equations. Extensive use is made of the properties of a Toeplitz covariance matrix. The Toeplitz property is a consequence of the covariance stationarity of the processes, and leads to a number of simplifications in the statistical inference.

Chapter 5 illustrates the application of spectral techniques to an econometric problem. Income and consumption time series are examined using their sample spectra, cross spectrum, and the adaptive distributed lag model described in Chap. 4.

2 Covariance Stationary Processes

2.1. Definition of a Stochastic Process

The theory of stochastic processes may be conveniently introduced as an extension of the elementary theory of probability. An experiment in which the outcome is subject to uncertainty may be described by a model with entities Ω, $\mathfrak{F}$, P—Ω being the space of sample points or outcomes of the experiment; $\mathfrak{F}$, the field of events generated by the sample points; and P, the probability measure defined on the events in the field $\mathfrak{F}$. According to a given rule, we assign to each outcome ω a number $X(\omega)$ that corresponds to an event in the field $\mathfrak{F}_x$. The measurable function

$$X = \{X(\omega), \ \omega \in \Omega\}$$

is called a *random variable* and P determines the probability distribution of X.

In the theory of stochastic processes, we describe an experiment by an analogous model in which we assign to each outcome ω a numerical-valued function $\{X(t, \omega); \ t \in T | \omega\}$. If t denotes time and T is the set of real numbers, then $\{X(t, \omega); \ t \in T | \omega\}$ is a function of time. For a given t, $X(t, \omega)$ is a number if ω is fixed, and is a random variable if ω varies. If ω and t both vary, then

$$X = \{X(t, \omega); \quad t \in T, \ \omega \in \Omega\}$$

7

is called a random or a *stochastic process*.† It is common practice in the literature to suppress the sample point argument ω so that, for fixed t and variable ω, $X(t)$ denotes a random variable. We may then denote the stochastic process by $\{X(t), \ t \in T\}$, but we shall generally be even more concise and denote it simply by X.

Two interpretations of a stochastic process are possible, depending on our choice of where to place emphasis. From the sample space point of view, we regard X as a collection of time functions indexed on ω. Each time function is a *realization* or a *sample function* of the stochastic process. From the time-domain point of view, X is a collection of random variables indexed on t. For our purposes, the latter interpretation is more convenient since the passage of time is an apparent reality and a collection of random variables indexed on time is an appealing model for describing the behavior of a phenomenon developing in time. This interpretation is also helpful when we actually observe a stochastic process over time. Here the indexed collection of observations is called a *time series* or a *sample record* of the stochastic process.

In elementary probability theory, we make probability statements about a random variable. When considering several random variables, we characterize them by the probability law governing their joint behavior. The determination of this joint behavior is one of the principal problems of multivariate statistical analysis wherein we regard the several random variables as forming a *vector random variable*. Since X is a family of random variables, the sequence $[X(t_1), X(t_2), \ldots, X(t_n)]$ is a vector random variable whose behavior is described by its n-dimensional probability distribution for any positive integer n. Making probability statements about the behavior of the stochastic process X at a set of n arbitrarily chosen points in time is, therefore, an exercise in multivariate statistical analysis.

The behavior of a random variable is completely described by its probability distribution; for a stochastic process, complete specification is more elaborate. Firstly, there is the probability distribution function (p.d.f.),

$$F(x, t) = P[X(t) \leq x],$$

that describes the behavior of the random variable $X(t)$. Secondly, for every pair of random variables $X(t_1)$ and $X(t_2)$, we have

$$F(x_1, x_2; t_1, t_2) = P[X(t_1) \leq x_1, X(t_2) \leq x_2].$$

† The sample space approach to defining a stochastic process may be found in detail in Rosenblatt [81, pp. 91–96]. Stochastic processes may also be defined as functions of spatial coordinates as in Bartlett [5, pp. 191–192].

Thirdly, for the vector random variable $[X(t_1), X(t_2), \ldots, X(t_n)]$, we have

(2.1) $F(x_1, x_2, \ldots, x_n; t_1, t_2, \ldots, t_n)$
$$= P[X(t_1) \le x_1, X(t_2) \le x_2, \ldots, X(t_n) \le x_n].$$

To describe completely the behavior of a stochastic process, we must know the joint probability distribution in Expression (2.1) for all $t_1, t_2, \ldots, t_n$ in T and for all positive integers n. The joint probability distribution must satisfy certain consistency conditions. The *symmetry* condition requires

$$F(x_{j_1}, x_{j_2}, \ldots, x_{j_n}; t_{j_1}, t_{j_2}, \ldots, t_{j_n})$$
$$= F(x_1, x_2, \ldots, x_n; t_1, t_2, \ldots, t_n)$$

where the sequence $j_1, j_2, \ldots, j_n$ is any permutation of the original indices $1, 2, \ldots, n$. This condition implies that the order in which the random variables are listed is immaterial. The *compatibility* condition requires that

$$F(x_1, x_2, \ldots, x_m, \infty, \ldots, \infty; t_1, t_2, \ldots, t_m, t_{m+1}, \ldots, t_n)$$
$$= F(x_1, x_2, \ldots, x_m; t_1, t_2, \ldots, t_m)$$

for any $t_{m+1}, \ldots, t_n$ if m is less than n.†

2.2. Strict Stationarity

So far we have made no assumption about how the random variables in the sequence of vector random variables are related. The nature of their association depends on the particular phenomenon we wish to describe. For many purposes it is convenient to consider two broad classes of stochastic processes, those that depend on historical time and those that do not. Stochastic processes in the former class are called *evolutionary*. Those in the latter class are known as *strictly stationary* processes; some writers define them as *stationary in the narrow sense* or *completely stationary*.

If the vector random variables $[X(t_1), X(t_2), \ldots, X(t_n)]$ and $[X(t_1 + s), X(t_2 + s), \ldots, X(t_n + s)]$ have the same n-dimensional p.d.f. for all values of the variable s, and all positive integer values of the index n, the stochastic process X is strictly stationary. This property implies that the joint p.d.f. depends on the time intervals between $t_1, t_2, \ldots, t_n$ and not on the time values themselves. The probability law governing the stochastic process is therefore invariant under time translation or, in other words, is independent of historical time.

---

† See, for example, Yaglom [106, p. 10] or Rosenblatt [81, p. 92].

Since most economic time-varying phenomena exhibit nonstationary behavior, it seems appropriate to classify them as evolutionary processes and to dismiss stationary models as inadequate representations. Doing so would be premature for two reasons. Firstly, we may relax the time invariance restriction to permit evolutionary behavior in a certain class of stationary models. This is done in Section 2.4. Secondly, we may perform transformations on economic time series that create new time series more stationary in appearance than the original ones. Techniques for accomplishing this transformation are discussed in Section 3.11.

2.3. The Autocovariance and Autocorrelation Functions

The mean and variance are the most common parameters used to describe the behavior of a random variable when its probability law is unknown. The mean measures location whereas the variance measures dispersion in the mean-square sense. They are computationally simple to estimate from sample data and have the added advantage of completely specifying the behavior of a normally distributed random variable.

In the theory of stochastic processes, the mean,

$$\mu(t) = E[X(t)] \equiv \int_{-\infty}^{\infty} \xi \, dF(\xi, t),$$

and autocovariance function,

$$R(s, t) = E\{[X(s) - \mu(s)][X(t) - \mu(t)]\}$$
$$\equiv \int_{-\infty}^{\infty} \int_{-\infty}^{\infty} \{[\xi_s - \mu(s)][\xi_t - \mu(t)]\} \, dF(\xi_s, \xi_t; s, t),$$

are used to describe behavior. Since

$$E\{[X(s) - \mu(s)]/[R(s, s)]^{\frac{1}{2}} - [X(t) - \mu(t)]/[R(t, t)]^{\frac{1}{2}}\}^2$$
$$= 2 - 2R(s, t)/[R(s, s)R(t, t)]^{\frac{1}{2}} \geq 0,$$

we have

$$R(s, t) \leq [R(s, s)R(t, t)]^{\frac{1}{2}}.$$

The mean function measures location for all t. The autocovariance function serves two purposes. Firstly, the variance function is given by

$$\text{var } [X(t)] = R(t, t),$$

and measures dispersion in the mean-square sense. Secondly, the autocovariance function conveys information about the *linear association* or *product-moment correlation* of the random variable $X(t)$ with the past and future behavior of the process. We refer to this relationship between the random variables composing the process as autocorrelation. It is measured by the autocorrelation function

$$\rho(s, t) = R(s, t)/[R(s, s)R(t, t)]^{\frac{1}{2}}$$
$$\rho(s, s) = 1; \ |\rho(s, t)| \leq 1.$$

If no correlation exists between the random variables $X(s)$ and $X(t)$, then

$$\rho(s, t) = \begin{cases} 1 & s = t \\ 0 & s \neq t. \end{cases}$$

If $X(s)$ and $X(t)$ are correlated, then $\rho(s, t)$ measures the degree of correlation. The rate at which the autocorrelation function ρ goes to zero as the interval $|s - t|$ increases indicates the extent of auto-correlation in the process.

If the vector random variable $[X(t_1), X(t_2), \ldots, X(t_n)]$ has a multivariate normal distribution for all $t_1, t_2, \ldots, t_n$ in T, then the specification of the mean and autocovariance functions suffices to describe completely the probability law governing the stochastic process X. In this case the process is called a *normal process*.

2.4. Covariance Stationarity

The definition of strict stationarity implies that

$$\mu = E[X(t)],$$
$$R(s - t) = E\{[X(s) - \mu][X(t) - \mu]\}.$$

That is, the mean is a constant and R is a function of one variable, namely the interval $(s - t)$. Neither the mean nor the autocovariance function is a function of historical time. This definition implies that for real-valued stochastic processes

$$R(0) > 0, \qquad R(\tau) = R(-\tau), \qquad \rho(\tau) = \rho(-\tau).$$

Also for any real sequence $\{a_j, j = 1, 2, \ldots, n\}$ and any n,

$$\sum_{j,k=1}^{n} a_j a_k R(s_j - t_k) = E\left\{\sum_{j=1}^{n} a_j[X(t_j) - \mu]\right\}^2 \geq 0.$$

A function with this property is said to be *positive semidefinite* or *nonnegative definite*. A function R for which strict inequality holds is said to be *positive definite*.

The purpose of our analysis is to make inferences about the probability laws governing time-varying phenomena. Because of the inherent difficulties in describing their joint probability distributions, most analytic methods concentrate on the mean and autocovariance functions. The concept of a strictly stationary process helps us in one way, but is unduly restrictive in another. Regarding the mean as a constant and the autocovariance function as a function of the separation $(s - t)$ simplifies statistical inference considerably. Requiring the historical time invariance of the probability law, however, restricts the class of phenomena that we may reasonably consider. Since our analysis is generally limited to the first- and second-order moments, there is no reason to dismiss processes whose higher-order moments are functions of historical time. To avoid this restriction, we define a class of processes that is broader than the strictly stationary class but retains the latter's requirements for the first- and second-order moments.

A stochastic process is said to be *covariance stationary* if

$$\mu = E[X(t)] < \infty$$
$$R(t, t + \tau) = R(\tau) = E[X(t + \tau)X(t)] - \mu^2 < \infty,$$
$$-\infty \leq t, \tau \leq \infty.$$

Some writers refer to processes with these properties as being *second-order*, *wide-sense*, or *mean-square stationary*. Covariance stationarity requires the autocorrelation to be a function of the time separation between random variables that constitute the process. The higher-order moments are unrestricted, and consequently the class includes all strictly stationary processes and may also include evolutionary processes. Notice that any normal process that is covariance stationary is also strictly stationary.

While it is true that covariance stationarity seldom if ever holds literally for time-varying phenomena, the concept does permit us to perform statistical analyses from which we can draw reasonably strong inferences about reality. As already mentioned, a phenomenon may often be made to conform more closely to stationarity by a suitable transformation.

2.5. The Ergodic Property

For a stochastic process to describe adequately the behavior of a time-varying phenomenon, certain parameters of the process must be known. Since a stochastic process, by definition, is tied to a scientific experiment, we may draw inferences about the parameters from sample records or time series for successive trials performed in the experiment. The greater the number of trials, the more sample records we collect, and the more stable our parameter estimates are.

An economic time series, however, comprises the data of one trial. We cannot recreate the same experimental environment in order to obtain more sample records. Since we have one time series, we have one observation on the phenomenon. Therefore, it would appear that the stability or reliability of our estimates is restricted by the impossibility of replicating the experiment. Fortunately this is not so. If we can establish certain properties for a stochastic process then we can derive statistically consistent estimates of its parameters from a single time series. A process with these properties is said to be *ergodic*. Since our interest is restricted to the mean and autocovariance function, we need only define ergodicity to the extent necessary for deriving consistent estimates of them from a single time series. Establishing the ergodicity of a process is beyond the scope of this study.

We earlier defined the mean μ and the autocovariance function R as probability averages over sample paths at given fixed times. We now consider the *time averages*

$$\hat{\mu}_T = T^{-1} \int_0^T X(t)\, dt$$

$$\hat{R}_T(\tau) = T^{-1} \int_0^T \{[X(t + \tau) - \mu][X(t) - \mu]\}\, dt$$

on the time path $(0, T)$. The mean-square error of $\hat{\mu}_T$ is

$$E(\hat{\mu}_T - \mu)^2 = E\left[T^{-1} \int_0^T X(t)\, dt - \mu\right]^2$$

$$= T^{-2} \int_0^T \int_0^T R(s - t)\, ds\, dt.$$

Letting $\tau = s - t$, we have

$$E(\hat{\mu}_T - \mu)^2 = T^{-2} \int_0^T \left[\int_0^{T-t} R(\tau)\, d\tau\right] dt$$

$$= T^{-1} \int_{-T}^T (1 - |\tau|/T)R(\tau)\, d\tau.$$

If

$$\lim_{T \to \infty} TE(\hat{\mu}_T - \mu)^2 = \int_{-\infty}^{\infty} R(\tau)\, d\tau,$$

then for large T the mean-square error is

$$T^{-1} \int_{-\infty}^{\infty} R(\tau) \, d\tau.$$

More generally, if

(2.2a) $$\lim_{T \to \infty} T^{-1} \int_{-T}^{T} R(\tau) \, d\tau = 0,$$

then $\hat{\mu}_T$ converges to μ in mean square. This in turn implies convergence in probability. That is, $\hat{\mu}_T$ is a consistent estimator of μ. We see that this convergence is accomplished using a single time series, that is, a single observation on the stochastic process. If Expression (2.2a) holds, we speak of the processes as being ergodic in the mean.

With regard to the sample autocovariance function, it can be shown that if

(2.2b) $$\lim_{T \to \infty} T^{-1} \int_{0}^{T} E\{[X(t + \tau + v)X(t + \tau) - R(v) - \mu^2]$$

$$\times [X(t + v)X(t) - R(v) - \mu^2]\} \, d\tau = 0,$$

then the time average $\hat{R}_\tau$ converges in mean square, and consequently in probability, to R_τ. Then $\hat{R}_\tau$ is a consistent estimator of R_τ. The truth of this assertion is demonstrated in Chap. 3 for a discrete process in the derivation of the sampling properties of its sample autocovariance function. For (2.2b) to hold we require the fourth-order moment function of X to be finite and a function of τ only [106, pp. 20–21].

Several points of clarification about ergodic, normal, and covariance stationary processes are worth making. For a normal process to be ergodic Expressions (2.2a) and (2.2b) must hold, the latter expression reducing to

(2.2c) $$\lim_{T \to \infty} T^{-1} \int_{0}^{T} R^2(\tau) \, d\tau = 0.$$

While covariance stationary processes have many advantages, they are not all ergodic. The following example illustrates this point. Consider a stochastic process defined by

(2.3) $$X(t) = a \cos \lambda_1 t + b \sin \lambda_1 t,$$

where a and b are independent normally distributed random variables with zero means and common variance σ^2. The autocovariance function of the process X is then

$$R(\tau) = \sigma^2 \cos \lambda_1 \tau.$$

We also have

$$A(T) = T^{-1} \int_0^T R(\tau) \, d\tau = \frac{\sigma^2 \sin \lambda_1 T}{\lambda_1 T},$$

$$B(T) = T^{-1} \int_0^T R^2(\tau) \, d\tau = \sigma^4 \left(\frac{1}{2} + \frac{\sin 2\lambda_1 T}{4\lambda_1 T} \right),$$

$$\lim_{T \to \infty} A(T) = 0, \qquad \lim_{T \to \infty} B(T) = \sigma^4 / 2.$$

Here ergodicity holds for the mean but not for the autocovariance function.

This process nicely illustrates an intuitively useful notion of ergodicity. Notice that remote sections of a given realization exhibit a high degree of dependence. In fact, observations that are separated by a time interval $2\pi/\lambda_1$ are perfectly correlated. As a result of this dependence, no additional information about the autocovariance function is gained by increasing T, the sample record length. Since we cannot reduce $B(T)$ by increasing T, the process is not ergodic with regard to the autocovariance function.

If a covariance stationary process has an autocovariance function such that

$$\lim_{T \to \infty} R(\tau) = 0,$$

then autocorrelation "wears off" as separation τ increases. For a normal process with this property, remote segments of a given realization are essentially independent so that, as T increases, we gain new information about the process. Here the process is ergodic, for the new information that accrues from continually lengthening the sample record permits us to derive consistent estimates of the parameters of the process.

2.6. Harmonic Analysis

To arrive at a proper idea of the value of covariance stationary processes for the analysis of economic time series, one must examine the stochastic model in Expression (2.3) in more detail. This course may seem odd to the reader since we have just described its inadequacy in conveying information about its autocovariance function. This model and closely related ones, however, have been used in economic analysis for over fifty years, beginning with Moore [64], and are therefore of some historical interest. A more important consideration is the fundamental link the model provides between the autocorrelated process X and a family of orthogonal random variables through which one may study X in detail unimpaired by autocorrelation. The truth of this assertion will slowly emerge as we proceed.

We begin by considering the stochastic process

(2.4) $$X(t) = a \cos \lambda_1 t + b \sin \lambda_1 t,$$

where a and b are random variables with

$$E(a) = E(b) = 0, \qquad E(a^2) = \sigma_a^2, \qquad E(b)^2 = \sigma_b^2.$$

No assumption is initially made about a common variance or orthogonality. We may reformulate Expression (2.4) as

$$a \cos \lambda_1 t + b \sin \lambda_1 t = ce^{i\lambda_1 t} + c^* e^{-i\lambda_1 t}$$
$$c = (a - ib)/2, \qquad c^* = (a + ib)/2.$$

The corresponding covariance function is

$$R(s, t) = E\{c^2 e^{i\lambda_1(s+t)} + cc^*[e^{i\lambda_1(s-t)} + e^{-i\lambda_1(s-t)}] + c^{*2} e^{-i\lambda_1(s+t)}\}.$$

For the process to be covariance stationary, the autocovariance function must be a function of $(s - t)$ only. That is, $R(s, t) = R(s - t)$, which requires that $E(c^2) = E(c^{*2}) = 0$. This occurs if

(2.5) $$E(a^2) = E(b^2), \qquad E(ab) = 0,$$

so that the random variables a and b are orthogonal and have a common variance. Were the random variables normally distributed, then orthogonality would imply independence. We therefore see that the process in Expression (2.3) is defined so as to insure covariance stationarity.

Now consider a process of the form

(2.6) $$X(t) = \sum_{j=1}^{n} (a_j \cos \lambda_j t + b_j \sin \lambda_j t)$$

$$= \sum_{j=1}^{n} (c_j e^{i\lambda_j t} + c_j^* e^{-i\lambda_j t}),$$

where

$$c_j = (a_j - ib_j)/2, \qquad c_j^* = (a_j + ib_j)/2$$
$$E(a_j) = E(b_j) = 0, \qquad E(a_j^2) = E(b_j^2) = \sigma_j^2 \qquad j = 1, 2, \ldots, n.$$

The corresponding autocovariance function is

$$R(s, t) = \sum_{j=1}^{n} \sum_{k=1}^{n} E[c_j c_k e^{i(\lambda_j t + \lambda_k s)} + c_j^* c_k e^{-i(\lambda_j t - \lambda_k s)}$$
$$+ c_j c_k^* e^{i(\lambda_j t - \lambda_k s)} + c_j^* c_k^* e^{-i(\lambda_j t + \lambda_k s)}].$$

In addition to the properties described in Expression (2.5), covariance stationarity now requires that for $j \neq k$

$$E(c_j c_k) = E(c_j^* c_k^*) = E(c_j^* c_k) = E(c_j c_k^*) = 0,$$

which implies that

$$E(a_j a_k - b_j b_k) = E(a_j a_k + b_j b_k) = 0$$
$$E(a_j b_k - a_k b_j) = E(a_j b_k + a_k b_j) = 0,$$

so that

(2.7) $$E(a_j a_k) = E(b_j b_k) = E(a_j b_k) = 0.$$

This result, together with that of Expression (2.5), means that the elements of each of the vector random variables

$$\mathbf{a} = [a_1, a_2, \ldots, a_n]$$
$$\mathbf{b} = [b_1, b_2, \ldots, b_n]$$
$$\mathbf{c} = [c_1, c_2, \ldots, c_n]$$

are orthogonal in pairs and, in addition, every element of $\mathbf{a}$ is orthogonal to every element in $\mathbf{b}$.

Imposing these requirements for covariance stationarity, the autocovariance function becomes

$$R(\tau) = \sum_{j=1}^{n} \sigma_j^2 \cos \lambda_j \tau \qquad \tau = s - t,$$

from which we observe that the variance of the process is an unweighted summation of the variances of the individual periodic components

(2.8) $$R(0) = \sum_{j=1}^{n} \sigma_j^2.$$

This representation of the variance has a number of features that promote a clear understanding of the nature of mean-square variation in the process. The additivity of the variance components and the absence of covariance terms permit us to determine the relative importance of each periodic component in the mean-square variation. The orthogonality of the components allows us to consider the mean-square variation as being generated by independent sources when the vector random variables **a** and **b** are normally distributed. At this point we begin to see that, although a process is autocorrelated, we may still consider its variance decomposition in terms of the variances of its component parts, if the process can be represented as in (2.6). Here, however, we have made no assumption about the higher-order moments of the vector random variables and therefore we regard only the component parts as orthogonal (except when they are normally distributed). For second-order moments, the interpretation of the source of mean-square variation is nevertheless the same.

The stochastic process in Expression (2.6) and its corresponding variance decomposition did not suddenly appear as fully developed tools of statistical analysis. In the late eighteenth and early nineteenth centuries, Lagrange, Euler, and Fourier developed the idea that an analytic function could be represented over a specified interval by a series of sines and cosines.† This representation is today a fundamental part of Fourier analysis in mathematics. Variance decomposition was a natural consequence of their work and was commonly called the harmonic analysis of a function. In 1898, Sir Arthur Schuster [85] suggested a way to perform statistical analysis on observed data in order to uncover the periodic components therein. He conceived of a phenomenon that could be described by Expression (2.6) plus a purely random additive disturbance. He suggested calculating the *periodogram*, which would serve as an estimate of the variance decomposition and thereby permit one to determine the relative importance of different periodic components. The periodogram is essentially a graph of the variances associated with the frequencies λ_j, $j = 1, 2, \ldots, n$ that are used in the analysis. His method was later refined by a number of writers.‡

The value of the periodogram for economics was first realized by Moore [64], who gave an account of Fourier analysis along with a periodogram analysis of rainfall in the Ohio Valley. The most notable analysis using this method in economics was performed by Beveridge [8], who studied wheat prices in Western Europe from 1545 to 1845. Figure 1 shows this periodogram.

As experience with periodogram analysis was gained during the years following Schuster's contribution, it became clear that this

---

† For a more complete historical description see Lanczos [58].
‡ See Davis [19] for references.

Source: Beveridge [8].

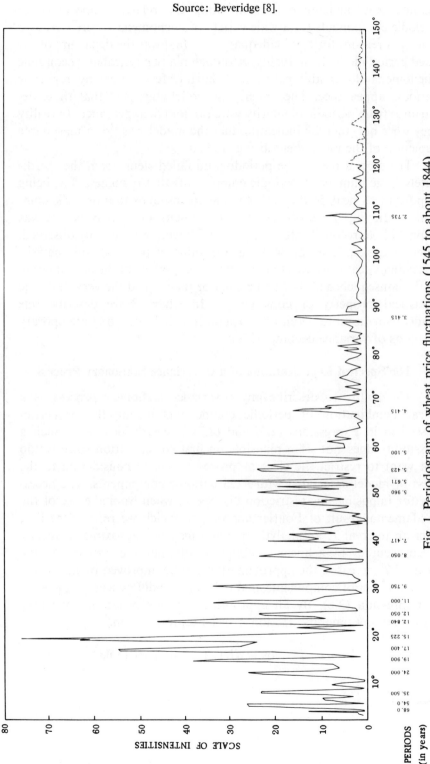

Fig. 1 Periodogram of wheat price fluctuations (1545 to about 1844)

method was inadequate for estimating the relative importance of periodic components for a wide variety of phenomena. There were two principal reasons for this inadequacy. The first was the departure of the fixed period model from reality. Although many physical and economic phenomena do exhibit recurrent behavior, few show any regularly periodic appearance. The underlying model suggested that these departures from periodic regularity were random in appearance. In reality they were not, thereby indicating that the model was not a reasonable depiction of the phenomena being studied.

The second reason the periodogram failed stems from the inordinately large number of periodic components that it suggested as being important. A peak in the periodogram corresponds to a periodic component. In Fig. 1 we observe the multiplicity of such peaks. It was impossible to reconcile all these peaks with what was actually observed. Therefore, skepticism arose about the value of periodogram analysis. Later analytical work was to explain this choppiness of the periodogram as the consequence of using an estimator that lacked the very desirable statistical property of consistency.† In Chap. 3 we describe this inadequacy in more detail and explain how it bears on contemporary methods of variance decomposition.

2.7. The Spectral Representation of a Covariance Stationary Process

The ability to describe any covariance stationary process as a linear combination of periodic components having the properties described in Expressions (2.5) and (2.7) is clearly desirable. Such a possibility does not seem plausible on first consideration since we do not want to restrict the class of processes we are considering to the class of model associated with fixed periodic components. The necessity for imposing this restriction disappears when we make use of the fundamental result of Fourier analysis, to which we referred earlier: over an interval T, an analytic function may be approximated to any given accuracy by a linear combination of sines and cosines. As the interval T increases, the approximation can be improved by increasing n, the number of components, and reducing the differences $(\lambda_{j+1} - \lambda_j)$ in Expression (2.6). By an appropriate limiting procedure one may show that any random function X has a representation‡

$$(2.9) \qquad X(t) = \int_0^\infty [\cos \lambda t \, dU(\lambda) + \sin \lambda t \, dV(\lambda)].$$

---

† See Bartlett [5, p. 278].
‡ See Yaglom [106, p. 36].

Expression (2.9) is a Fourier-Stieltjes integral and, for our purposes, U and V are stochastic processes indexed on the continuous parameter λ. Now taking X to have a zero mean, its autocovariance function is

$$R(s, t) = \int_0^\infty \int_0^\infty \{\cos \lambda t \cos \omega s E[dU(\lambda)\, dU(\lambda)]$$
$$+ \sin \lambda t \sin \omega s E[dV(\lambda)\, dV(\omega)]$$
$$+ \sin \lambda t \cos \omega s E[dV(\lambda)\, dU(\omega)]$$
$$+ \cos \lambda t \sin \omega s E[dU(\lambda)\, dV(\omega)]\}.$$

In order for X to be covariance stationary, R must be a function of $t - s$. Covariance stationarity therefore requires that

$$E[dU(\lambda)\, dU(\omega)] = E[dV(\lambda)\, dV(\omega)] = \begin{cases} 2\, dG(\lambda) & \lambda = \omega \\ 0 & \lambda \neq \omega \end{cases}$$

$$E[dU(\lambda)\, dV(\omega)] = 0 \quad \text{all } \lambda, \omega.$$

The function G is called the cumulative spectrum; its significance will be explained shortly.

It is instructive first to consider several properties of the processes U and V. If

$$E\{[U(\lambda_1) - U(\lambda_2)][U(\lambda_3) - U(\lambda_4)]\} = 0$$

for any nonoverlapping closed intervals, $[\lambda_2, \lambda_1]$ and $[\lambda_4, \lambda_3]$, we say that U is a stochastic process with uncorrelated nonoverlapping increments and is called an orthogonal process. As the nonoverlapping intervals become infinitesimally small, we may, using an appropriate limiting procedure, replace the differenced quantities $[U(\lambda_j) - U(\lambda_{j+1})]$ $[V(\lambda_j) - V(\lambda_{j+1})]$ by $dU(\lambda)$ and $dV(\lambda)$ respectively.† Notice that if U and V are orthogonal processes then X is covariance stationary.

It is convenient to write

$$X(t) = \tfrac{1}{2} \int_{-\infty}^\infty \{e^{i\lambda t}[dU(\lambda) - i\, dV(\lambda)] + e^{-i\lambda t}[dU(\lambda) + i\, dV(\lambda)]\},$$

where

$$dU(\lambda) \equiv dU(-\lambda), \qquad dV(\lambda) \equiv -dV(-\lambda).$$

---

† See Yaglom [106, pp. 36–43] for a more complete description.

Then

(2.10)
$$X(t) = \int_{-\infty}^{\infty} e^{i\lambda t} \, dZ(\lambda),$$

$$dZ(\lambda) = [dU(\lambda) - i \, dV(\lambda)]/2,$$

$$dZ^*(\lambda) = [dU(\lambda) + i \, dV(\lambda)]/2,$$

$$E[dZ(\lambda) \, dZ(\omega)] = \begin{cases} dG(\lambda) & \lambda = -\omega \\ 0 & \lambda \neq -\omega, \end{cases}$$

$$E[dZ(\lambda) \, dZ^*(\omega)] = \begin{cases} dG(\lambda) & \lambda = \omega \\ 0 & \lambda \neq \omega. \end{cases}$$

The convenience of this representation is immediately apparent, for we note that

(2.11) $\quad R(s - t) = E[X(s)X(t)] = \displaystyle\int_{-\infty}^{\infty} \int_{-\infty}^{\infty} e^{i(\lambda s + \omega t)} E[dZ(\lambda) \, dZ(\omega)]$

$$= \int_{-\infty}^{\infty} e^{i\lambda(s-t)} \, dG(\lambda),$$

(2.12)
$$\rho(s - t) = \int_{-\infty}^{\infty} e^{i\lambda(s-t)} \, dF(\lambda),$$

$$dF(\lambda) = dG(\lambda)/R(0),$$

$$F(-\infty) = 0, \qquad F(\infty) = 1,$$

F being the spectral distribution function.

We may now express the variance of X as

$$R(0) = \int_{-\infty}^{\infty} dG(\lambda),$$

so that $dG(\lambda)$ is a contribution to the variance made by an infinitesimal increment $dZ(\lambda)$. The relative importance of this increment in the variance is $dF(\lambda)$. By analogy with the interpretation of **c**, the variance X may be regarded as made up of additive contributions of variance attributable to uncorrelated nonoverlapping increments that form the orthogonal process Z.

Expression (2.12) is the essence of the Wiener-Khintchine theorem that forms the basis for the generalized harmonic analysis of covariance stationary processes [102, 54]. The theorem states that any stationary process has an autocorrelation function representable as in Expression (2.12) and, if an autocorrelation function may be represented as in Expression (2.12), there exists a stationary process with this autocorrelation function. This result follows from a more general theorem of Bochner stating that, in order for F to be a distribution function, ρ must be a positive semidefinite function.† The positive semidefinite character of R and hence ρ has already been established in Section 2.4. The converse of this theorem also holds so that, if F is a distribution function, then ρ is a positive semidefinite function.

It can be shown that if

$$(2.13) \qquad \lim_{\tau \to \infty} \rho(\tau) = 0,$$

then the spectral distribution function F is absolutely continuous. The condition in Expression (2.13) implies the absence of regular periodic elements in X, for, as we have shown, such an element has a periodic autocorrelation function. It can in fact be shown that if F is discontinuous, the points of discontinuity correspond to the periodic elements in X. The condition in (2.13) is a reasonable one, for it implies that the influence of the past wears off as time elapses. In the remainder of the book we assume Expression (2.13) holds unless the contrary is stated. As a further implication we note that a covariance stationary normal process with a continuous spectrum is ergodic since Expressions (2.2a) and (2.2b) hold.

Since F is an absolutely continuous function we may write

$$dF(\lambda) = f(\lambda)\,d\lambda, \qquad dG(\lambda) = g(\lambda)\,d\lambda,$$

where f and g are the spectral density function and the spectrum respectively. The spectrum is the basis used to describe the properties of covariance stationary processes in this study. Since f and g form Fourier transform pairs with ρ and R, respectively, we may express their inverse transforms as

$$(2.14) \qquad f(\lambda) = (2\pi)^{-1} \int_{-\infty}^{\infty} e^{-i\lambda\tau} \rho(\tau)\,d\tau,$$

$$(2.15) \qquad g(\lambda) = (2\pi)^{-1} \int_{-\infty}^{\infty} e^{-i\lambda\tau} R(\tau)\,d\tau.$$

It can also be shown that the absence of regularly periodic components in X implies that U, V, and therefore Z are also absolutely continuous so that we may write

$$dU(\lambda) = u(\lambda)\,d\lambda, \qquad dV(\lambda) = v(\lambda)\,d\lambda, \qquad dZ(\lambda) = z(\lambda)\,d\lambda.$$

It may often be convenient to use this notation in deriving results.

---

† See Yaglom [106, p. 47].

2.8. Relationship between Variance Decomposition and Frequency

So far we have referred to λ_n as a real-valued index only. Looking back at Expression (2.4), we note that the larger is λ_1, the greater is the *frequency* of regular oscillation per unit time. In economics, it is customary to describe an oscillation by its period N, the length of time required for one complete oscillation. For current purposes it is more convenient to describe an oscillation by its frequency N^{-1}, measured in cycles per unit time. The quantity λ is simply *angular frequency*, $\lambda = 2\pi/N$, which is the number of revolutions around the unit circle per unit time.

Observe that for the model in Expression (2.6) the quantity

$$\sigma_j^2 \bigg/ \sum_{k=1}^{n} \sigma_k^2$$

is the proportion of variance attributable to the component with frequency λ_j. Now, the larger is λ_j, the more rapid is the variation attributable to the jth component in the process per unit time. Since n components are present, however, the relative importance of the mean-square variation per unit time of the jth component depends on the proportion of variance attributable to it. If the proportion is small, then, regardless of the value of λ_j, this component will not substantially influence the appearance of the process. The crucial point of interest is that a process may contain a number of frequency components that separately have widely varying appearances. When brought together, as in Expression (2.6), the proportion of variance attributable to each component determines the extent to which it influences the pattern of mean-square variation of the process.

Several differences are to be stressed when one turns to the spectral representation in Expression (2.11). We are now considering a continuum of angular frequencies so that λ takes on all values on the real line. For this reason, we can no longer allude to the variance contribution of a particular frequency. We now refer to the contribution of a small band of frequencies around a particular λ. It is not possible for a single frequency to dominate the variation, but it is possible for a band of frequencies to dominate. Several examples in Section 2.11 will illustrate this point.

Looking at Expression (2.11), we observe that each random variable $dZ(\lambda)$ is associated with a band of frequencies. Since these variables are uncorrelated for nonoverlapping intervals, the variance of the process is the integral summation of the variance attributable to each $dZ(\lambda)$. Suppose we wish to determine the extent of mean-square variation over some time interval T^*. If the proportion $[1 - F(2\pi/T^*) + F(-2\pi/T^*)]$ were relatively large, we would expect to observe considerable variation over the interval T^* relative to longer intervals. This is a consequence of the variance being primarily associated with oscillations whose periods are shorter than T^*. Were the proportion $[1 - F(2\pi/T^{**}) + F(-2\pi/T^{**})]$ close to zero for T^{**} considerably greater than T^*, then we would expect to observe little variation over the interval T^*. In this case the variation is attributable to oscillations whose periods are considerably larger than T^*.

From this description we may conclude that studying a covariance stationary process in the frequency domain permits us to conceptualize mean-square variation in terms of a continuum of periodic components. In addition, it permits us to work with orthogonal processes U, V, and Z that are considerably simpler to work with than is the autocorrelated process X in the time domain. *When working with orthogonal processes, one need only consider their variances. There are no covariance considerations as with correlated random variables.*

2.9. Decomposition of the Spectral Distribution Function

In assuming the function F to be absolutely continuous, we preclude the presence of certain phenomena in the underlying stochastic process. Before we proceed, it will be worthwhile to explain the justification for our assumption. Since F is a distribution function, it may be represented by the classical decomposition:

$$F(\lambda) \equiv F_1(\lambda) + F_2(\lambda) + F_3(\lambda),$$

where F_1, F_2 and F_3 are nondecreasing functions of angular frequency. The first component F_1 is an absolutely continuous function with a derivative f. The second component F_2 is composed of a finite number of saltuses or steps. For every saltus, there exists a corresponding perfectly periodic function with the same frequency in the stochastic process. The remaining component F_3 is the continuous singular component of the function F.

The component F_3 does not appear to have any practical significance, and we consequently omit further mention of it. To assume the presence of the step function F_2 would also be unrealistic, since it is unlikely that many if any economic processes contain perfectly periodic components. While it is true that the seasonal components of economic time series do display considerable regularity, careful analysis shows that these components differ from year to year in amplitude and phase and are, therefore, not perfectly periodic. For this reason, we may safely assume the absence of any saltuses in the function F. This leaves us with the absolutely continuous function F_1, which is precisely the component on which our exposition has concentrated.

2.10. Relationship between the Autocorrelation and Spectral Density Functions

Since the autocorrelation function and the spectral density function form a Fourier transform pair, knowledge of one implies a knowledge of the other. This means that both functions contain exactly the same information about intertemporal dependence. The reader may therefore be somewhat confused by the introduction of and emphasis on the spectral density function rather than the more conventional autocorrelation function. There are good reasons, however, for preferring to study the properties of the process $\{X(t)\}$ in the frequency rather than in the time domain. Apart from the simpler sampling properties that Chap. 3 describes, the spectral density function provides a clearer understanding of what constitutes a process than the autocorrelation function does.

The autocorrelation function stresses dependence along the time axis. The spectral density function conveys this same information in the squared amplitude associated with oscillations at different frequencies. That is, it describes the process in terms of the relative importance of different kinds of oscillations that shape it. Characterizing a process in terms of uncorrelated additive contributions permits greater ease of analysis than does correlation analysis where each value of the autocorrelation function is the weighted summation of the same contributions.

It is to be noted that the spectral distribution function bears the same relationship to the autocorrelation function that a probability distribution does to a characteristic function. In probability theory, knowledge of the probability distribution function implies knowledge of its characteristic function, and vice versa, so that both functions contain the same information about the probability law governing the corresponding random variable. Why then do we more often stress the former than the latter? The probability distribution function is additive, and has a clearly understood meaning, integrates to unity, and lends itself to easy graphical interpretation. The characteristic function is not additive in the same way and has no easily understood graphical interpretation. The desirability of these properties is responsible for the more common use of the probability distribution function. The spectral density function also has these properties, whereas the autocorrelation function does not. The former is consequently easier to interpret than the latter.

Behavioral relationships in economics are generally specified in temporal, not frequency, terms. Questions of interest concern how the effects of a change in an economic variable are distributed in time and what the time delay is between a change induced in one variable by a change in another. It is quite natural to ask, then, what conceptual assistance spectral methods offer when studying time-varying economic phenomena. The remaining sections of this chapter attempt to answer this question.

2.11. Examples

The autocorrelation and spectral density functions for three elementary covariance stationary processes are listed in Table 1. In the *first* example, the autocorrelation function is a negative exponential, and the spectral density function has the form of the Cauchy distribution. The corresponding process fluctuates irregularly about its mean, and its reluctance to change value (that is, its degree of autocorrelation) is inversely related to the parameter α. This reluctance is evident in

Table 1

Three Autocorrelation Function and Spectral Density Function Pairs

No.	Autocorrelation Function	Spectral Density Function
1	$e^{-\alpha|\tau|}$	$\dfrac{\alpha}{\pi(\alpha^2 + \lambda^2)}$
2	$e^{-\alpha|\tau|}\cos\lambda_1\tau$	$\dfrac{\alpha}{2\pi}\left[\dfrac{1}{\alpha^2 + (\lambda_1 - \lambda)^2} + \dfrac{1}{\alpha^2 + (\lambda_1 + \lambda)^2}\right]$
3	$.5e^{-\alpha|\tau|}(\cos\lambda_1\tau + \cos\lambda_2\tau)$	$\dfrac{\alpha}{4\pi}\left\{\displaystyle\sum_{i=1}^{2}\left[\dfrac{1}{\alpha^2 + (\lambda_i - \lambda)^2} + \dfrac{1}{\alpha^2 + (\lambda_i + \lambda)^2}\right]\right\}$

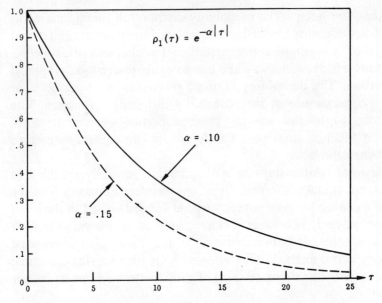

Fig. 2 Autocorrelation function (first example)

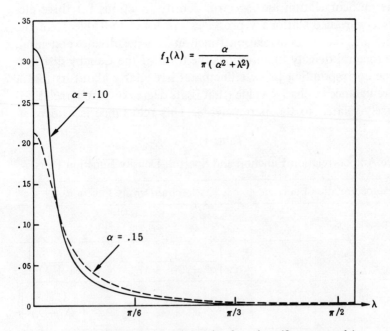

Fig. 3 Spectral density function (first example)

Fig. 2, where the smaller α is associated with a less damped auto-correlation function.

In Fig. 3, the more autocorrelated process (smaller α) has a greater concentration of variance in the low-frequency range than the less autocorrelated process does. At high frequencies its spectral density function is lower than that for the less autocorrelated process. This implies that, if both processes have the same variance, a change in value of a given magnitude will occur over a longer time interval in the more autocorrelated process. Alternatively, the more autocorre-lated process changes less in value in a given time interval. Notice also that no single frequency dominates the variance in either process. The concentration of variance occurs in the low-frequency range.

In the *second* example, the autocorrelation function is again damped, but now oscillates about zero as shown in Fig. 4. Were we to observe the corresponding process, some semblance of periodicity would be apparent. The spectral density functions in Fig. 5 show peaks in the vicinity of the angular frequency $\pi/6$. In the more autocorrelated process (smaller α), the peak is greater in magnitude and narrower in width, indicating a more pronounced and regular periodicity than in the less autocorrelated process. If time were measured in months, then the angular frequency $\pi/6$ would be the fundamental seasonal fre-quency corresponding to a period of twelve months. The model of the second example would then depict a sinusoidal seasonal component whose amplitude and phase are changing slowly compared to the fundamental seasonal period of twelve months; the more rapid is the rate of change, the more obscured is the seasonal component and, therefore, the less sharp is the peak in the spectral density function.

From this example, the reader may further develop his under-standing of the distinction between correlation and spectral analyses. In order to describe the *period* of oscillation from Fig. 4, we measure the length of a complete oscillation on the time axis. In Fig. 5, we may choose the peak midband frequency to describe the oscillation.

In the *third* example we consider a stochastic process containing a seasonal oscillation with two frequency components whose amplitudes and phase angles are slowly varying in time. We observe in Fig. 6 that there is a six-month as well as a twelve-month cycle. We cannot tell from Fig. 6, however, what the relative importance of each of the two seasonal components is. We can tell their relative importance using the spectral density function in Fig. 7. Here we observe that the peaks associated with both frequencies have approximately equal magnitudes

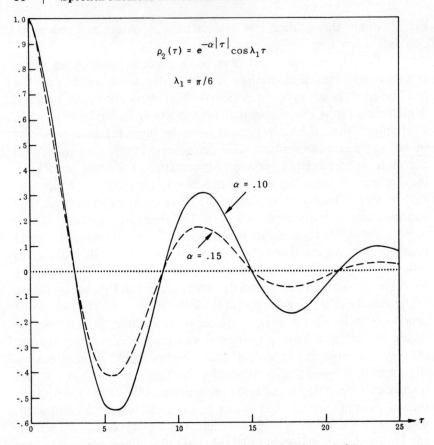

Fig. 4 Autocorrelation function (second example)

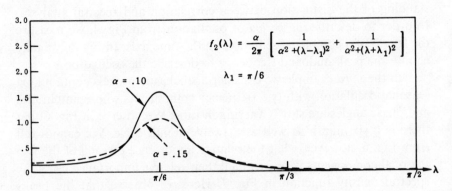

Fig. 5 Spectral density function (second example)

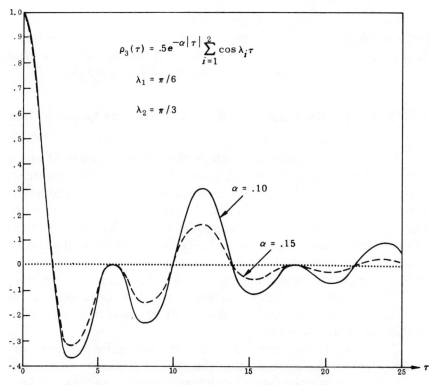

Fig. 6 Autocorrelation function (third example)

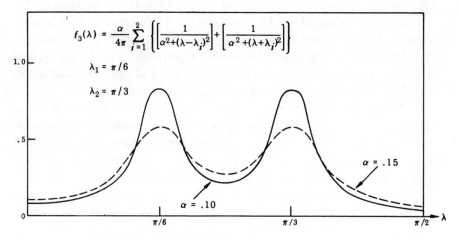

Fig. 7 Spectral density function (third example)

and widths. The ability to discern the relative importance of different frequency bands is quite valuable in our understanding of the variation in a process. A graph of the autocorrelation function does not yield this information; a graph of the spectral density function does.

2.12. Spectral Methods and a Traditional Model of an Economic Time Series

Time-varying phenomena in economics are often described as the summation of four independent contributions. They are called the trend, cyclical, seasonal, and irregular components. Sometimes a multiplicative model is assumed, in which case the summation is in terms of the logarithmic transformations of these components. The trend component represents the evolutionary change that occurs in economic phenomena over long time intervals. It emerges primarily as a changing mean level around which the remaining components fluctuate with different degrees of regularity. Trend also appears in the magnitude of the fluctuations around this mean level and may be considered as a change in their covariance structure.

The cyclical component describes the successive advance and declines, other than seasonal, that characterize so much of economic behavior. An economic phenomenon may contain more than one kind of cycle and, because of their superposition, it may be difficult to observe either cycle by visual inspection of the time series. If these cycles exhibit sufficient regularity in their respective periods of oscillation, we would observe peaks and concentrations of variance in the vicinity of their corresponding frequencies. The narrower each peak is relative to its width, the more regular and discernible the cycle is. The advantage of the spectral approach in this case is that each cycle is identified separately along the frequency axis, and the superposition in the time domain is no longer an impediment to identification.

The cyclical component is often so irregular that the corresponding spectrum shows only a concentration of variance over the whole low-frequency range. The absence of a peak does not mean the absence of a cyclical phenomenon. Indeed, all covariance stationary autocorrelated phenomena are cyclical, but very often their periods of oscillation are highly irregular. A peak in the spectrum identifies a reasonably regular cycle; the absence of a peak does not preclude the presence of an irregular cycle. Kendall [52] and more recently Howrey [46] have discussed several ways of defining cycle length. Howrey has also attempted to show how these several definitions relate to a peak in the spectrum.

The seasonal component represents the observed within-year pattern that is superimposed on economic phenomena by climatic and institutional factors. It is considerably more regular in appearance than the cyclical component, and its spectrum has peaks and concentrations of variance at some or all of the seasonal frequencies, $\pi j/6$, $j = \pm 1$, $\pm 2, \ldots, \pm \infty$. It is very common for the seasonal component to slowly change character from year to year so that one may speak of the presence of a seasonal trend. In this case the widths of the peaks indicate the rate of change of the seasonal elements relative to the seasonal frequencies.[†] Broad rounded peaks indicate that the seasonal pattern is changing rapidly, whereas narrow peaks imply a relatively slow-changing pattern. Even though changes in the seasonal pattern may be trend-like, it is still possible to speak of the seasonal component as being covariance stationary.

The remaining irregular component represents the presence of purely random phenomena. This component has no observable pattern and, as Section 2.17 will show, has a uniform spectrum. This follows logically since no particular frequency dominates. If one frequency did, then the irregular component would show some regularity or cyclical behavior, which is impossible by definition.

Figure 8 shows the sample spectra of manufacturers' monthly shipments of durable goods, seasonally unadjusted and adjusted.[‡] The concentrations of variance in the low-frequency ranges of the two spectra are similar, both being due to the trend and cyclical components. The remaining peaks in the spectrum of the unadjusted series occur at the seasonal frequencies, the peak at the fundamental seasonal frequency $\pi/6$ being obscured by the low-frequency concentration. Notice that these peaks do not appear in the spectrum of the adjusted series, indicating the absence of the seasonal component.

The presence of trend in economic phenomena implies a dependence on historical time and, consequently, a literal violation of the assumptions of stationarity. The possibility of structural changes in the behavior of the other three components also contributes to our skepticism about the use of stationary models. To circumvent these difficulties, we distinguish trend as related to a changing mean from

---

[†] This concept is discussed in more detail in Davenport and Root [18, pp. 158–159].
[‡] The time series was taken from Ref. 93, pp. 25 and 214, and Ref. 92, Vol. 47, No. 2, February 1967, and No. 4, April 1967.

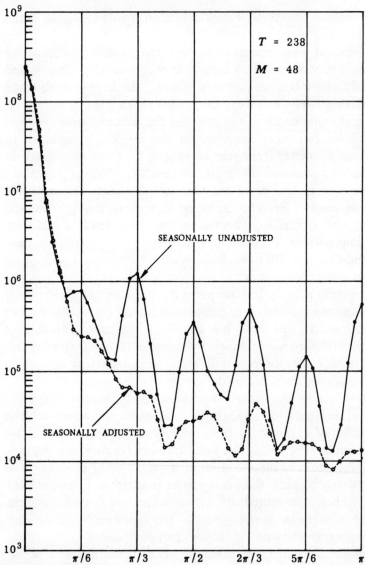

Fig. 8 Sample spectra of manufacturers' shipments of durable goods, 1947–1966

evolutionary modifications in the other three components, which are considered changes in the covariance structure. To begin with, one should note that, over any finite time interval, trend in the mean cannot be distinguished from low-frequency contributions that shape economic phenomena. There is indeed no a priori reason why we cannot consider trend to be immersed in the low-frequency content of economic phenomena. If we accept this broad interpretation, then we may speak of a trend-cycle component, a characterization which is in common use.

The stationarity assumptions preclude a changing variance as well as a changing mean. Over short periods of time economic phenomena appear homoscedastic; therefore, there is no major problem. Over long time intervals, however, the fluctuations in these phenomena grow in absolute magnitude with growth in the mean. If the change in magnitude is significant, a logarithmic transformation of the sample data may be used to amplify fluctuations at the low points and attenuate them at the high points. The result is to give a more uniform appearance to fluctuations and, therefore, a more stable variance. This is especially true if the variance of the original process is a linear function of its mean. In Section 3.11 we describe other methods of making a time series more stationary.

One of the principal aims of economic time series analysis is to decompose a time series into trend, cyclical (nonseasonal), seasonal, and irregular components in order to study them individually. The initial approach of spectrum analysis is to look at a frequency decomposition, remembering the association between certain components and frequencies. In this way it is possible to study carefully the effects of attempted removals of any of the four components on the time series as a whole.† Section 2.27 demonstrates one such approach related to seasonal adjustment.

In applying spectral methods to the analysis of an economic process, we assume that a large number of elementary cycles with different frequencies have contributed to the shaping of the fluctuations in the process. The spectral density function tells us the average relative contribution made by a small band of these elementary cycles to the mean-square variation. Since period is inversely proportional to frequency, we are actually measuring the average relative contribution of elementary cycles with different periods. If an economic cycle does show a strong regular tendency, its spectral density function will show a concomitant peak at a corresponding frequency and a concentration of variance in the neighborhood of this peak. This, of course, was the case in Section 2.11 for the example that contained seasonal components.

---

† See, for example, Nerlove [67, 68].

2.13. Aliasing

Economic phenomena are generally observed periodically. While prices on stock and commodity exchanges are recorded continuously, most data collection methods restrict observation to daily, weekly, monthly, or yearly intervals. An economic time series is therefore usually the result of sampling a phenomenon at equally spaced points in time. The periodic sampling procedure creates a problem in correctly identifying the sources of mean-square variation in the observed phenomenon. The problem is known as *aliasing* and a thorough understanding of its implications is necessary before we can appropriately interpret estimated spectra.

Let Δt be the time interval between successive observations. Assuming there are no errors of observation, our model now becomes a stochastic sequence

$$X_t = X(t) \qquad t = 0, \pm \Delta t, \pm 2 \, \Delta t, \ldots, \pm \infty$$

with autocovariance function

$$R_\tau = R(\tau) = \int_{-\infty}^{\infty} g(\lambda) e^{i\lambda\tau} \, d\lambda, \qquad \tau = 0, \pm \Delta t, \pm 2 \, \Delta t, \ldots, \pm \infty.$$

We may also represent the autocovariance function as

$$R_\tau = \sum_{j=-\infty}^{\infty} \int_{(2j-1)\pi/\Delta t}^{(2j+1)\pi/\Delta t} g(\lambda) e^{i\lambda\tau} \, d\lambda$$

$$= \int_{-\pi/\Delta t}^{\pi/\Delta t} g(\lambda) e^{i\lambda\tau} \, d\lambda + \int_{\pi/\Delta t}^{3\pi/\Delta t} g(\lambda) e^{i\lambda\tau} \, d\lambda + \int_{3\pi/\Delta t}^{5\pi/\Delta t} g(\lambda) e^{i\lambda\tau} \, d\lambda$$

$$+ \cdots + \int_{-3\pi/\Delta t}^{-\pi/\Delta t} g(\lambda) e^{i\lambda\tau} \, d\lambda + \int_{-5\pi/\Delta t}^{-3\pi/\Delta t} g(\lambda) e^{i\lambda\tau} \, d\lambda + \cdots$$

$$= \int_{-\pi/\Delta t}^{\pi/\Delta t} [g(\lambda) e^{i\lambda\tau} + g(\lambda + 2\pi/\Delta t) e^{i(\lambda+2\pi/\Delta t)\tau}$$

$$+ g(\lambda + 4\pi/\Delta t) e^{i(\lambda+4\pi/\Delta t)\tau} + \cdots$$

$$+ g(\lambda - 2\pi/\Delta t) e^{i(\lambda-2\pi/\Delta t)\tau}$$

$$+ g(\lambda - 4\pi/\Delta t) e^{i(\lambda-4\pi/\Delta t)\tau} + \cdots] \, d\lambda.$$

Since τ is measured in multiples of Δt, we have

$$R_\tau = \int_{-\pi/\Delta t}^{\pi/\Delta t} g_a(\lambda) e^{i\lambda\tau} \, d\lambda$$

(2.16) $\qquad g_a(\lambda) = \sum_{j=-\infty}^{\infty} g(\lambda + 2j\pi/\Delta t) \qquad |\lambda| \leq \pi/\Delta t,$

g being the spectrum of $\{X(t)\}$, and g_a the spectrum of $\{X_t\}$.

Expression (2.16) implies that the sequence $\{X_t\}$ has a spectrum g_a that is restricted to the interval $(-\pi/\Delta t, \pi/\Delta t)$. In addition, g_a does not coincide with g. The spectrum of $\{X_t\}$ is actually a confounding of variance contributions of the continuous process $\{X(t)\}$. The confounding of frequency contributions is known as *aliasing* and creates an ambiguity in determining the origins of mean-square variation in $\{X(t)\}$. Figure 9 will help the reader visualize the meaning of aliasing. If observations are made periodically at intervals Δt, then the observer cannot distinguish between the sinusoid with period $\pi/\Delta t$ and that with period $2\pi/\Delta t$.

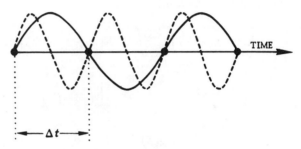

Fig. 9 Aliasing

If observations are made at intervals Δt, then we can estimate the spectrum only for the interval $(-\pi/\Delta t, \pi/\Delta t)$, where $\pi/\Delta t$ is known as the *Nyquist frequency*. This frequency corresponds to a period of $2\,\Delta t$, and consequently we lose all information about fluctuations with shorter periods. If these fluctuations are significant, g_a will be a poor approximation to g in the interval $(-\pi/\Delta t, \pi/\Delta t)$. If the spectrum g vanishes beyond this interval, then there is no ambiguity or loss of information. No general statement can be made, however, about variation in economic phenomena without at least specifying the sampling interval Δt. If a phenomenon is observed monthly and strong weekly fluctuations are present, then an aliasing problem will exist. Similarly, quarterly observations may create an aliasing problem since fluctuations with periods less than six months will be confounded with the true spectrum in the estimation interval.

2.14. Unweighted Time Averaging

Aliasing is a serious problem which, if neglected, can easily lead to erroneous conclusions. The method of observation in economics may, however, reduce the importance of aliasing. Economic time series are often averages. Production, for example, is measured in units produced over some finite time interval. Consumption, income, and investment are similarly measured. We may therefore consider our observation to be a sequence,

$$X_t = (\Delta t)^{-1} \int_{t-\Delta t}^{t} X(s)\, ds = (\Delta t)^{-1} \int_{-\Delta t}^{0} X(t+s)\, ds,$$

$$t = 0, \pm\Delta t, \pm 2\,\Delta t, \ldots, \pm\infty,$$

where Δt is the averaging as well as the sampling interval. The autocovariance function is

$$R_\tau = (\Delta t)^{-2} \int_{-\Delta t}^{0} \int_{-\Delta t}^{0} R(\tau + s_1 - s_2)\, ds_1\, ds_2$$

$$= (\Delta t)^{-2} \int_{-\infty}^{\infty} g(\lambda) e^{i\lambda\tau} \left[\int_{-\Delta t}^{0} \int_{-\Delta t}^{0} e^{i\lambda(s_1 - s_2)}\, ds_1\, ds_2 \right] d\lambda$$

$$= \int_{-\infty}^{\infty} A(\lambda) g(\lambda) e^{i\lambda\tau}\, d\lambda,$$

where

$$A(\lambda) = [2 \sin (\lambda\, \Delta t/2)/(\lambda\, \Delta t)]^2,$$

and the spectrum of the sequence is

$$h(\lambda) = \sum_{j=-\infty}^{\infty} A(\lambda + 2j\pi/\Delta t) g(\lambda + 2j\pi/\Delta t) \qquad |\lambda| \leq \pi/\Delta t.$$

As Fig. 10 shows, the function A attenuates each frequency contribution, especially those greater than the Nyquist frequency π. If the spectrum g is relatively small in magnitude above this frequency, then the *folding back* of frequency components induces a small degree of aliasing in the principal frequency interval. Working with unweighted averages, therefore, reduces the importance of the aliasing problem. It does not eliminate it altogether, however, especially if these exists a strong source of mean-square variation above the Nyquist frequency.

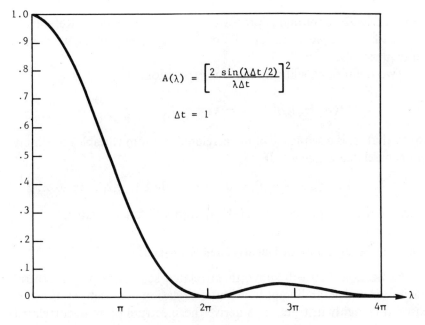

$$A(\lambda) = \left[\frac{2 \sin(\lambda \Delta t/2)}{\lambda \Delta t}\right]^2$$

$$\Delta t = 1$$

Fig. 10 The attenuation function

2.15. Covariance Stationary Sequences

In the remainder of this study, we consider the covariance stationary stochastic sequence $\{X_t;\ t = 0, \pm 1, \pm 2, \ldots, \pm \infty\}$ with mean, autocovariance, autocorrelation, and spectral density functions:

$$\mu = E(X_t), \qquad R_\tau = E(X_t X_{t+\tau}) - \mu^2, \qquad \rho_\tau = R_\tau/R_0,$$

$$R_\tau = \int_{-\pi}^{\pi} e^{i\lambda\tau} g(\lambda)\, d\lambda, \qquad \rho_\tau = \int_{-\pi}^{\pi} e^{i\lambda\tau} f(\lambda)\, d\lambda,$$

$$g(\lambda) = (2\pi)^{-1} \sum_{\tau=-\infty}^{\infty} R_\tau e^{-i\lambda\tau},$$

$$f(\lambda) = (2\pi)^{-1} \sum_{\tau=-\infty}^{\infty} \rho_\tau e^{-i\lambda\tau} \qquad -\pi \leq \lambda \leq \pi.$$

Again the functions R, ρ, g, and f are symmetric around zero. It is also to be noted that X_t has a spectral decomposition as in Expression (2.11), the limits of integration now being $\pm\pi$.

Since observations are usually made periodically, it is convenient for us to develop spectral concepts for stochastic sequences to which we shall sometimes refer as discrete processes. A completely analogous theory exists for continuous processes and may be found in References 81 and 105. Since we hereafter refer exclusively to sequences, we denote the sequence $\{X_t\}$ by X.

Two definitions will simplify our exposition. If

$$E[(X_t - \mu_x)(X_{t+\tau} - \mu_x)] = \begin{cases} R_{x,0} & \tau = 0 \\ 0 & \tau \neq 0, \end{cases}$$

we say that X, is a sequence of uncorrelated random variables or simply an uncorrelated sequence. If

$$E[(X_t - \mu_x)(Y_{t+\tau} - \mu_y)] = 0 \qquad \tau = 0, \pm 1, \pm 2, \ldots, \pm \infty,$$

we say that the sequences X and Y are mutually uncorrelated.

2.16. The Spectrum of an Uncorrelated Sequence

As Section 2.20 will show, uncorrelated sequences play a central role in our understanding of how autocorrelation materializes. For this reason it is highly instructive to derive the spectrum of an uncorrelated sequence prior to describing how transformations of the sequence can produce autocorrelation.

Let $\{\epsilon_t\}$ be an uncorrelated sequence with mean zero and variance σ^2. Its autocovariance function is then

$$R_{\epsilon,\tau} = E(\epsilon_t \epsilon_{t+\tau}) = \begin{cases} \sigma^2 & \tau = 0 \\ 0 & \tau \neq 0, \end{cases}$$

and its spectrum is

$$g_\epsilon(\lambda) = (2\pi)^{-1} \sum_{\tau=-\infty}^{\infty} E(\epsilon_t \epsilon_{t+\tau}) e^{-i\lambda\tau}$$

$$= \sigma^2/(2\pi) \qquad -\pi \leq \lambda \leq \pi.$$

This means that the spectrum of an uncorrelated sequence is uniform over the domain of g_ϵ and, consequently, all frequencies in $(-\pi, \pi)$ contribute equally to the variance. This result is not surprising, since an uncorrelated sequence has no noticeable pattern of mean-square variation through time and, hence, the contributions due to low frequencies should be the same as those due to high frequencies.

Consider a stochastic sequence $\{\epsilon_t\}$ whose elements are mutually independent and identically distributed. The sequence is obviously strictly stationary and, since it is a special kind of uncorrelated sequence, it has a uniform spectrum. We refer to such a sequence as *white noise*.

2.17. Linear Filtering

We have noted that an observable economic time-varying phenomenon is, in reality, the superposition of a number of more fundamental phenomena, each of whose behavior we cannot directly observe without removing the remaining ones. We may, for example, wish to remove seasonal effects in order to study the remaining seasonally adjusted time series. In another case, smoothing out short term fluctuations may be desirable in order to observe the trend component more closely. Still another modification may attempt to remove all but the random component from a series in order to determine its relative importance as measured by a comparison of variances before and after removal.

These modifications are often accomplished by performing linear transformations or *filtering* on the original data. As we are about to show, the effects of linear filtering can be easily determined in the frequency domain. Consider a covariance stochastic sequence $\tilde{X}$ such that

$$(2.17a) \qquad \tilde{X}_t = \sum_{s=-p}^{q} a_s X_{t-s} \qquad \sum_{s=-p}^{q} a_s^2 < \infty,$$

where the filtering sequence $\{a_s\}$ is real valued, and p and q are positive constants. The autocovariance function of $\tilde{X}$ is

$$(2.17b) \qquad \tilde{R}_\tau = \sum_{r,s=-p}^{q} a_r a_s R_{\tau+r-s},$$

which is a double moving average on the autocovariance function R. Because of the averaging in Expressions (2.17a) and (2.17b), it is difficult to discern in advance the effects of linear filtering in the time domain. Taking Fourier transforms on both sides of Expression (2.17b), we have

$$\tilde{g}(\lambda) = (2\pi)^{-1} \sum_{\tau=-\infty}^{\infty} \tilde{R}_\tau e^{-i\lambda\tau}$$

$$= (2\pi)^{-1} \sum_{\tau=-\infty}^{\infty} \sum_{r,s=-p}^{q} a_r a_s R_{\tau+r-s} e^{-i\lambda\tau}$$

$$= (2\pi)^{-1} \sum_{r=-p}^{q} a_r e^{i\lambda r} \sum_{s=-p}^{q} a_s e^{-i\lambda s} \sum_{\tau=-\infty}^{\infty} R_\tau e^{-i\lambda\tau}$$

$$= |A(\lambda)|^2 g(\lambda),$$

where

$$A(\lambda) = \sum_{s=-p}^{q} a_s e^{-i\lambda s}$$

is called the *frequency response function* or the *transfer function*. Notice that averaging in the time domain corresponds to multiplication in the frequency domain. We are, therefore, able to determine the effect of averaging on the spectrum at each frequency by the amplification or attenuation attributable to $|A(\lambda)|^2$.

Replacing X_t in Expression (2.16) by its spectral representation, we have

$$(2.18) \qquad \tilde{X}_t = \sum_{s=-p}^{q} a_s \int_{-\pi}^{\pi} e^{i\lambda(t-s)} z(\lambda)\, d\lambda = \int_{-\pi}^{\pi} e^{i\lambda t} A(\lambda) z(\lambda)\, d\lambda.$$

The frequency response function A is usually complex valued, and for this reason it is often instructive to study Expression (2.18) as well as Expression (2.16). We may write

$$A(\lambda) = A_1(\lambda) + iA_2(\lambda) = G(\lambda) e^{i\phi(\lambda)},$$

where

$$G(\lambda) = [A_1^2(\lambda) + A_2^2(\lambda)]^{\frac{1}{2}},$$
$$\tan \phi(\lambda) = A_2(\lambda)/A_1(\lambda).$$

The functions G and ϕ are called the *gain* and *phase angle* respectively. Noting that

$$z(\lambda) = [u(\lambda) - iv(\lambda)]/2 = |z(\lambda)| e^{-i\theta(\lambda)}$$

and

$$\tan \theta(\lambda) = v(\lambda)/u(\lambda),$$

we may write

$$X_t = \int_{-\pi}^{\pi} e^{i[\lambda t - \theta(\lambda)]} |z(\lambda)|\, d\lambda$$

and

$$(2.19) \qquad \tilde{X}_t = \int_{-\pi}^{\pi} e^{i[\lambda t - \theta(\lambda) + \phi(\lambda)]} G(\lambda) |z(\lambda)|\, d\lambda.$$

We observe that a linear transformation involves two distinct modifications. One is an amplification or attenuation of each frequency component due to the gain $G(\lambda)$. The other is a phase shift at each frequency which is attributable to $\phi(\lambda)$. If low-frequency contributions are amplified and high-frequency ones are attenuated, the resulting sequence $\tilde{X}$ will appear to have less variation over short time intervals and more variation over long intervals. The effect would be to increase the extent of autocorrelation. If low-frequency contributions are attenuated and high-frequency ones amplified, the resulting sequence $\tilde{X}$ will be quite a bit choppier in appearance over short intervals and not so variable over long intervals. The extent of autocorrelation would be reduced in this case.

The interpretation of the phase angle is somewhat more difficult. In Expression (2.19), we observe that filtering introduces an additive phase shift at each frequency. The effect of this shift is to change the time relationships among the phases of the frequency contributions. This shift is generally undesirable because it may, for example, cause a business cycle in the filtered series to turn down before it actually does in the original series.

To illustrate the interpretation of phase shift, we consider an example wherein

(2.20) $$\phi(\lambda) = \lambda\nu,$$

so that

$$\tilde{X}_t = \int_{-\pi}^{\pi} e^{i\lambda(t+\nu)} G(\lambda) z(\lambda) \, d\lambda.$$

This permits us to write

$$\tilde{X}_t = \sum_{s=-p}^{q} b_s X_{t+\nu-s}.$$

Using Expression (2.16), we may show that

$$b_s = a_{s-\nu},$$

$$\sum_{s=-p}^{q} b_s \cos \lambda s = G(\lambda),$$

$$\sum_{s=-p}^{q} b_s \sin \lambda s = 0,$$

which occurs for

$$b_s = b_{-s}, \qquad p = q.$$

That is, $\tilde{X}_t$ is formed by a symmetric weighting and averaging around the time point $t + \nu$. If ν equals zero, then there is no phase shift and the symmetric averaging takes place around the point t. This form of averaging is commonly used in economic analysis. The next two sections contain examples illustrating its use. We observe that, while amplification or attenuation does occur, the time relationships among the frequency components is preserved if ν equals zero.[†]

There exists a fundamental ambiguity in the interpretation of $\phi(\lambda)$ as a negative or positive delay at frequency λ. To see this, we note that it is also correct to write

$$\phi(\lambda) \pm 2n\pi = \tan^{-1}\left[A_2(\lambda)/A_1(\lambda)\right] \qquad n = 0, 1, \ldots, \infty.$$

Heuristically speaking, if

$$X_t = \sin \lambda t$$

so that

$$\tilde{X}_t = \sin\left[\lambda t + \theta(\lambda) \pm 2n\pi\right],$$

then the time delay between X and $\tilde{X}$ is

$$(\theta(\lambda) \pm 2n\pi)/\lambda.$$

If, for example, λ were the seasonal frequency $\pi/12$ and the phase angle resulting from the linear filtering were $2\pi/3$, then the time delay would be $4 \pm 12n$ months, where n is indeterminate unless additional information is available. The reader is cautioned to remember that a negative phase angle does not necessarily imply $\tilde{X}$ lags X at frequency λ. In the remainder of the book, we adopt the convention of taking n to be zero unless additional information dictates otherwise.

We may briefly summarize the effects of linear filtering on the sequence X as follows: the gain induces amplification or attenuation of the amplitudes of the components of X whereas the phase angle induces a translation along the time axis of the phases of these components.

2.18. A Method of Seasonal Adjustment

In order to interpret trends and cyclical movements in economic phenomena, it is often necessary to "seasonally adjust" the time series of interest. Many adjustment procedures have been suggested and, in recent years, spectral methods have been used to study their effects.[‡]

---

† Jenkins [50] has conveniently reproduced graphs of gain and phase angle by Bode for a number of different types of linear filters.
‡ See, for example, Hannan [39, 40] and Nerlove [67, 68].

One naive method of seasonal adjustment is accomplished by a linear filter with

$$a_s = \tfrac{1}{12}, \qquad s = 0, \pm 1, \pm 2, \ldots, \pm 5$$

$$a_6 = a_{-6} = \tfrac{1}{24}.$$

This yields the adjusted series

$$\tilde{X}_t = \tfrac{1}{24}(X_{t+6} + X_{t-6}) + \tfrac{1}{12} \sum_{s=-5}^{5} X_{t-s}$$

for which we have

$$A(\lambda) = \frac{\sin 6\lambda \cos (\lambda/2)}{12 \sin (\lambda/2)}.$$

Since A is a real-valued function, there is no phase shift and we therefore need only consider the gain as shown in Fig. 11. Notice that this form of seasonal filtering has successfully eliminated all the seasonal frequency components, $\pi/6$, $\pi/3$, $\pi/2$, $2\pi/3$, $5\pi/6$ and π. In addition, the high-frequency contribution has been considerably attenuated. This is an additional virtue of the filter because it has effectively smoothed out many short-term irregularities that might otherwise be misleading. Notice that the filter has also substantially reduced the low-frequency contribution and has thereby smoothed out the very phenomenon of interest. This oversmoothing makes the filter inadequate for most purposes, a fact long apparent to economists without the aid of spectral methods. This example illustrates how the filter may change the appearance of an original time series to a considerably greater extent than is desirable. Inspection of the frequency response function permits one to observe this before linear filtering is performed.

2.19. The Long Swings Hypothesis

The existence of long swings in certain economic phenomena has been a topic of controversy for many years. In testing this hypothesis, economists customarily use yearly data which has been smoothed by a number of filtering operations. Kuznets [51] investigated the long swings hypothesis using data that had been subjected to two linear filters applied in series.† Before describing his approach, we will find it instructive to study the effects of applying filters in series.

---

† The spectrum analysis of Kuznets' investigation has been studied by Adelman [1] and Hatanaka and Howrey [44]. The subsequent description is based on Howrey [46].

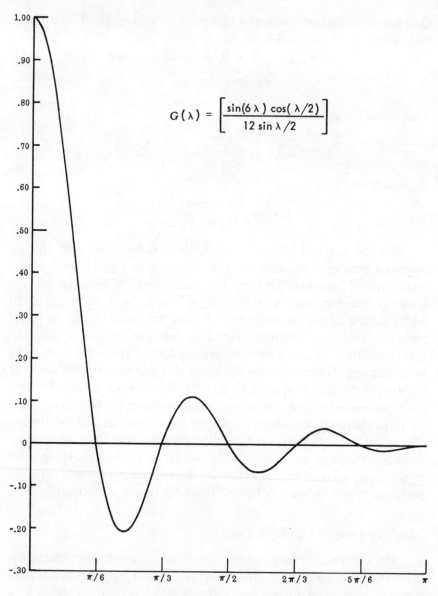

$$G(\lambda) = \left[\frac{\sin(6\lambda)\cos(\lambda/2)}{12\sin\lambda/2}\right]$$

Fig. 11 Gain of seasonal adjustment filter

Suppose the first filtering operation yields

$$\tilde{X}_t = \sum_{s=-p_1}^{q_1} a_s X_{t-s},$$

and the second yields

$$\tilde{\tilde{X}}_t = \sum_{r=-p_2}^{q_2} b_r \tilde{X}_{t-r} = \sum_{r=-p_2}^{q_2} \sum_{s=-p_1}^{q_1} b_r a_s X_{t-r-s}.$$

It follows directly that

$$\tilde{\tilde{R}}_\tau = \sum_{r_1, r_2 = -p_2}^{q_2} b_{r_1} b_{r_2} \tilde{R}_{\tau + r_1 - r_2}$$

$$= \sum_{r_1, r_2 = -p_2}^{q_2} \sum_{s_1, s_2 = -p_1}^{q_1} b_{r_1} b_{r_2} a_{s_1} a_{s_2} R_{\tau + r_1 - r_2 + s_1 - s_2},$$

$$\tilde{\tilde{g}}(\lambda) = |B(\lambda)|^2 \tilde{g}(\lambda) = |B(\lambda)|^2 |A(\lambda)|^2 g(\lambda),$$

where A and B are the frequency response functions corresponding to the first and second filtering operations respectively. One may show in general that the effect of passing a time series through a series of n filters is to multiply the original spectrum by the squared gains of the frequency response functions corresponding to the n filters. One may also show that the resulting phase shift is simply the summation of the individual phase shifts.

In the Kuznets analysis of long swings, the filtering procedure covered an eleven-year interval [57, p. 321]. The first filter was of the form

$$a_s = \tfrac{1}{5}, \qquad s = 0, \pm 1, \pm 2,$$

so that

$$\tilde{X}_t = \tfrac{1}{5} \sum_{s=-2}^{2} X_{t-s},$$

$$A(\lambda) = \frac{\sin (5\lambda/2)}{5 \sin (\lambda/2)}.$$

The second filter involved a differencing operation,

$$b_5 = 1, \qquad b_{-5} = -1,$$

so that

$$\tilde{\tilde{X}}_t = \tilde{X}_{t+5} - \tilde{X}_{t-5},$$

$$B(\lambda) = 2i \sin 5\lambda.$$

Notice that the first filter causes no phase shift whereas the second causes a shift of $\pi/2$ at all frequencies. Here we concern ourselves with

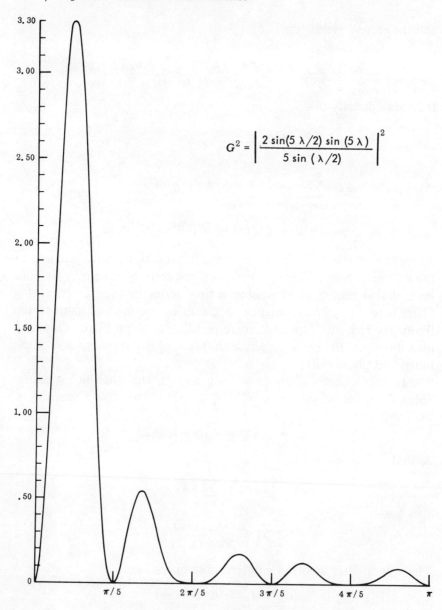

$$G^2 = \left| \frac{2 \sin(5\,\lambda/2) \sin(5\,\lambda)}{5 \sin(\lambda/2)} \right|^2$$

Fig. 12 Squared gain of Kuznets' adjustment filter

the shortcoming of this series filtering procedure due to its gain characteristic. Figure 12 shows the squared gain for the series filter. The diagram indicates that from the point of view of mean-square variation the filter essentially passes low-frequency contributions but virtually suppresses all high-frequency contributions.

This narrow band filter may in effect create a peak in the output spectrum $\tilde{\tilde{g}}$ even though none may exist in the original one. If the spectrum g were uniform, thus indicating an uncorrelated sequence, the spectrum $\tilde{\tilde{g}}$ would have a high peak in the low-frequency range as shown in Fig. 12. This frequency would correspond to a period in the time domain of approximately 20.26 years. It is interesting to note that Kuznets concluded that the average period of the long swings he observed was about 20 years [57, p. 320]. Even though Kuznets used a different method of estimating the average period, the similarity in conclusions makes one wonder if the regularity was perhaps induced by the series filtering operation.

Howrey [46] has pointed out that, even if a peak existed in the spectrum, the series filter used by Kuznets would shift its position on the frequency axis in the spectrum $\tilde{\tilde{g}}$. Were the true peak to the right of the peak low frequency (but less than $\pi/5$), the shift would be to the left. The effect of this shift would be to bias the data in favor of a longer period. If the peak were to the left of the low frequency peak, the shift would be to the right, thereby shortening the period of oscillation. It is also worthwhile noting that the frequency variance component in the original spectrum g at $19\pi/200$ is more than tripled in amplitude in the resulting spectrum $\tilde{\tilde{g}}$ by the series filter, thereby adding another note of skepticism.

2.20. The Moving Average Representation

The justification for using spectral methods to study economic phenomena stems from the presence of autocorrelation therein. In its absence, traditional statistical methods would suffice for meaningful analysis. It is somewhat surprising to learn that uncorrelated sequences do play a significant role in the presentation of economic behavior by stochastic sequences. Uncorrelated sequences form a basic building block in representing a wide class of covariance stationary sequences. Slutzky [88] and Yule [108] first noted this fact in their investigations, which show that a linear combination of uncorrelated random variables can generate cyclical behavior similar to that observed in economic time series.† More general and formal is the Wold decomposition theorem which states that every covariance stationary sequence can

---

† Granger and Hatanaka [30] describe the spectral implications of Slutzky's particular result.

be represented as†

$$(2.21a) \qquad X_t = V_t + W_t,$$

$$(2.21b) \qquad V_t = \sum_{j=1}^{n} b_j e^{i\lambda_j t},$$

$$(2.21c) \qquad W_t = \sum_{s=0}^{\infty} a_s \epsilon_{t-s},$$

where $\{a_s\}$ is a sequence of constants with

$$(2.21d) \qquad \sum_{s=0}^{\infty} a_s^2 < \infty,$$

$\{b_j\}$ and $\{\epsilon_t\}$ are sequences of uncorrelated random variables (the latter with mean zero and variance σ^2), and

$$\text{cov}\,(V_t, W_s) = 0.$$

We call $\{V_t\}$ a *deterministic* process and $\{W_t\}$ an *indeterministic* one.

This result is significant, for it implies that any covariance stationary sequence can be expressed as the sum of two mutually uncorrelated sequences—one of which is representable as a linear combination of sines and cosines with random coefficients, as described in Section 2.6, and the other of which is a one-sided *moving average* of a sequence of uncorrelated random variables with zero mean and common finite variance.

Since we have assumed that X has no periodic (deterministic) components, we need only consider the moving average representation.

$$(2.22) \qquad X_t = \sum_{s=0}^{\infty} a_s \epsilon_{t-s}.$$

If the elements of $\{\epsilon_t\}$ are mutually independent and identically distributed, X_t is called a *linear process*. For the moment, we need only consider $\{\epsilon_t\}$ to be an uncorrelated sequence. The autocovariance function of X is

$$(2.23) \qquad R_\tau = \sigma^2 \sum_{s=0}^{\infty} a_s a_{s+\tau}.$$

Expression (2.23) implies that the sequence X is autocorrelated.

---

† For a discussion and proof see Wold [105, pp. 84–89 and 198–200] or Cox and Miller [15, pp. 286–288].

The spectrum of the moving average is

$$g(\lambda) = \frac{\sigma^2}{2\pi} \sum_{\tau=-\infty}^{\infty} \sum_{s=0}^{\infty} a_s a_{s+\tau} e^{-i\lambda\tau}$$

$$= \frac{\sigma^2}{2\pi} |A(\lambda)|^2,$$

where

(2.24)
$$A(\lambda) = \sum_{s=0}^{\infty} a_s e^{-i\lambda s}.$$

The frequency response function A plays the same role here as described in Section 2.18. This is not surprising since the sequence $\{a_s\}$ is a linear filter. Here the uncorrelated sequence has a uniform spectrum, as expected, which is "shaped" by the frequency response function, thereby inducing autocorrelation.

If the elements of $\{\epsilon_t\}$ are mutually independent and identically distributed, it follows that this sequence is strictly stationary. An obvious question now arises as to the stationarity of X. The characteristic function of the vector random variable $[X_t, X_{t+1}, \ldots, X_{t+n}]$ is

$$\varphi(\phi) = E\{\exp[i(\phi_1 X_t + \phi_2 X_{t+1} + \cdots + \phi_{n+1} X_{t+n})]\}$$

$$= E\left\{\exp\left[i \sum_{s=0}^{\infty} (\phi_1 a_s \epsilon_{t-s} + \phi_2 a_s \epsilon_{t+1-s} + \cdots + \phi_{n+1} a_s \epsilon_{t+n-1})\right]\right\}$$

$$= E\left\{\exp\left[i \sum_{s=0}^{\infty} \epsilon_{t-s}(\phi_1 a_s + \phi_2 a_{s-1} + \cdots + \phi_{n+1} a_{s-n})\right]\right\}$$

$$= E\left\{\exp\left[i \sum_{s=0}^{\infty} \epsilon_{t-s}\phi_s\right]\right\},$$

where

$$a_s = 0 \qquad s < 0,$$

$$\phi_s = \phi_1 a_s + \phi_2 a_{s-1} + \cdots + \phi_{n+1} a_{s-n}.$$

Since the elements of $\{\epsilon_t\}$ are independent and identically distributed, we may write

(2.25)
$$\varphi(\phi) = \prod_{s=0}^{\infty} E[\exp(i\epsilon_{t-s}\phi_s)] = \prod_{s=0}^{\infty} \varphi_\epsilon(\phi_s),$$

where φ_ϵ is the characteristic function of an element in $\{\epsilon_t\}$. Since $\{\epsilon_t\}$ is strictly stationary, we conclude from Expression (2.25) that the linear process X is also strictly stationary.

2.21. The Linear Autoregressive Representation

Expression (2.22) provides a convenient way of relating an auto-correlated sequence $\{X_t\}$ to an uncorrelated sequence $\{\epsilon_t\}$. Now it is often desirable to express X_t in terms of its past values. This is especially true when our purpose is to devise a prediction model for some future value $X_{t+\nu}$. Consider the linear *autoregressive* model,

$$(2.26) \qquad \sum_{s=0}^{p} b_s X_{t-s} = \epsilon_t,$$

that permits us to represent X_t as a linear combination of past values plus a random term uncorrelated with the past. This representation was first studied by Yule [108].

For stationary sequences, there often exists a close relationship between the moving average in (2.22) and the autoregressive scheme in (2.26). Noting that

$$B(\lambda)\, dZ_x(\lambda) = dZ_\epsilon(\lambda),$$

$$B(\lambda) = \sum_{s=0}^{p} b_s e^{-i\lambda s},$$

we have the important result that

$$g_x(\lambda) = \frac{\sigma^2}{2\pi |B(\lambda)|^2},$$

so that $A(\lambda) = 1/B(\lambda)$, provided that the roots of the polynomial

$$\sum_{s=0}^{p} b_s z^s = 0$$

are outside the unit circle. For example, the sequence

$$X_t - \alpha X_{t-1} = \epsilon_t \qquad |\alpha| < 1$$

has both autoregressive and moving average representations, since the corresponding root $\alpha^{-1} > 1$. By contrast the sequence

$$X_t - X_{t-1} = \epsilon_t$$

has an autoregressive but not a moving average representation, since the corresponding root is on the unit circle. Since all covariance stationary sequences have a moving average representation, a sequence with an autoregressive scheme such that

$$\sum_{s=0}^{p} b_s z^s = 0$$

has at least one root on the unit circle is not covariance stationary. One further point of clarification is useful. If the power series

$$\sum_{s=0}^{\infty} a_s z^s = 0$$

has a zero on the unit circle, then X has a moving average but not an autoregressive representation. The case

$$X_t = \epsilon_t - \epsilon_{t-1}$$

is an example.

Consider next the mixed representation

$$\sum_{s=0}^{p} b_s X_{t-s} = \sum_{s=0}^{q} a_s \epsilon_{t-s},$$

so that

$$g_x(\lambda) = \frac{\sigma^2 |A(\lambda)|^2}{2\pi |B(\lambda)|^2},$$

provided that the roots of

$$\sum_{s=0}^{p} b_s z^s$$

are outside the unit circle. In this case X is called a sequence with a rational spectrum since g_x is the ratio of two polynomials in $e^{i\lambda}$.

Bartlett [5] and Hannan [36] describe estimating and testing procedures for autoregressive coefficients, and Whittle [101] explains the significance of autoregressive schemes for prediction theory.

There are several questions concerning the uniqueness of an autoregressive scheme that deserve attention. These questions are answered in Section 2.24, but first we shall discuss the details of two important specifications in Sections 2.22 and 2.23.

2.22. The First-Order Autoregressive Process

Suppose that X had an autoregressive representation

(2.27a) $$X_t = \alpha X_{t-1} + \epsilon_t \qquad |\alpha| < 1,$$

so that

$$b_s = \begin{cases} \alpha & s = 1 \\ 0 & s \neq 1. \end{cases}$$

Expression (2.27a) implies that X_t is a function of its most recent past and an uncorrelated additive disturbance. Observe that

(2.27b) $$B(\lambda) = 1 - \alpha e^{-i\lambda},$$

which implies that in the moving average representation

(2.27c) $$A(\lambda) = \sum_{s=0}^{\infty} \alpha^s e^{-i\lambda s} = (1 - \alpha e^{-i\lambda})^{-1}$$

$$a_s = \alpha^s,$$

$$X_t = \sum_{s=0}^{\infty} \alpha^s \epsilon_{t-s}.$$

The spectrum of X is

(2.27d) $$g(\lambda) = \sigma^2 [2\pi(1 - 2\alpha \cos \lambda + \alpha^2)]^{-1}.$$

If X_{t-1} and ϵ_t are independent, then X is called a *Markov process*.

Figure 13 shows the spectra corresponding to this sequence for two values of the parameter α. The variance σ^2 is taken to be unity. Observe that the more autocorrelated sequence has a higher concentration of variance at low frequencies than does the less autocorrelated sequence. The opposite situation holds in the high-frequency range. Since the variance of X is $\sigma^2/(1 - \alpha^2)$ we have the interesting fact that the high-frequency content of the less autocorrelated sequence is higher even though its total variance is less.

If α were in the interval $(-1, 0)$, the peak in the spectrum would be at angular frequency π. This implies a periodicity of two time units which would become more regular in shape as α approaches -1. We observe that the first-order autoregressive model permits two contrasting interpretations, depending on whether α is positive or negative.

2.23. The Second-Order Autoregressive Process

While the first-order autoregressive scheme is convenient for expository purposes, it often proves inadequate as a representation of economic behavior. The model precludes any concentration of variance at frequencies other than zero or $\pm\pi$. It is desirable to describe an economic process by a model that permits cyclical behavior with an arbitrary frequency and unrestricted degree of regularity. The second-order autoregressive scheme offers this flexibility.†

---

† Howrey [46] has used this model to study the effects of linear filtering on the shifting of spectral peaks. Kendall [52] has studied this model in detail and derived the weights in the moving average expression when the roots are complex.

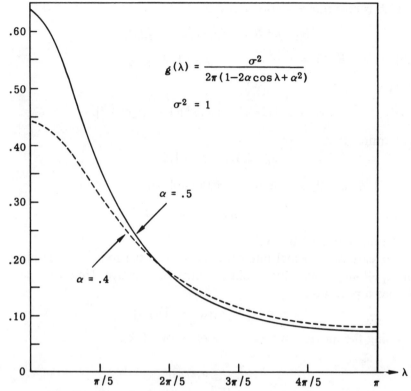

$$g(\lambda) = \frac{\sigma^2}{2\pi(1-2\alpha\cos\lambda+\alpha^2)}$$

$$\sigma^2 = 1$$

$\alpha = .5$

$\alpha = .4$

Fig. 13 Spectral density function: first-order Markov process

To derive this representation we make use of the series filtering concepts described in Section 2.19. The first linear filter yields

$$\tilde{X}_t = X_t - \beta_1 X_{t-1},$$

and the second,

$$\tilde{\tilde{X}}_t = \tilde{X}_t - \beta_2\tilde{X}_{t-1} = X_t - (\beta_1 + \beta_2)X_{t-1} + \beta_1\beta_2 X_{t-2}.$$

If the two series filtering operations transform X into a sequence of uncorrelated random variables,

$$\tilde{\tilde{X}}_t = \epsilon_t,$$

then we have the second-order autoregressive scheme

$$X_t - \alpha_1 X_{t-1} - \alpha_2 X_{t-2} = \epsilon_t$$
$$\alpha_1 = \beta_1 + \beta_2, \qquad \alpha_2 = -\beta_1\beta_2.$$

If X is covariance stationary, then

$$|B_1(\lambda)|^2 |B_2(\lambda)|^2 g(\lambda) = \sigma^2/(2\pi),$$

(2.28) $\qquad B_j(\lambda) = 1 - \beta_j e^{i\lambda} \qquad j = 1, 2,$

so that

(2.29) $\qquad g(\lambda) = \sigma^2/[2\pi|(1 - \beta_1 e^{i\lambda})(1 - \beta_2 e^{i\lambda})|^2].$

This requires that

$$|\beta_j| < 1 \qquad j = 1, 2.$$

That is, $1/\beta_1$ and $1/\beta_2$, the roots of the polynomial

$$(1 - \beta_1 z)(1 - \beta_2 z) = 0$$

must be outside the unit circle.

The case of principal interest occurs when the roots are complex. Then $\alpha_1^2 + 4\alpha_2 < 0$. Notice that α_2 must be negative. Here the spectrum has a peak at

$$\lambda_0 = \cos^{-1} [\alpha_1(\alpha_2 - 1)/\alpha_2].$$

Using the moving average representation of X_t,

$$X_t = \sum_{s=0}^{\infty} a_s \epsilon_{t-s},$$

we have

$$a_0 = 1 \qquad a_1 = \alpha_1$$

$$a_n - \alpha_1 a_{n-1} - \alpha_2 a_{n-2} = 0 \qquad n = 2, 3, \ldots, \infty.$$

The solution to this homogenous difference equation is

$$a_s = c_1 \beta_1^s + c_2 \beta_2^s, \qquad s > 0$$

since

$$\beta_1 = [\alpha_1 + (\alpha_1^2 + 4\alpha_2)^{\frac{1}{2}}]/2, \qquad \beta_2 = [\alpha_1 - (\alpha_1^2 + 4\alpha_2)^{\frac{1}{2}}]/2.$$

Using the definitions of a_0 and a_1, the general solution is

$$X_t = (\beta_1 - \beta_2)^{-1} \sum_{s=0}^{\infty} (\beta_1^{s+1} - \beta_2^{s+1}) \epsilon_{t-s}.$$

The second-order autoregressive model overcomes the inadequacy of the first-order model by permitting concentrations of variance at frequencies other than zero and $\pm\pi$. While this property is desirable in economic analysis, the second-order scheme may still not adequately represent a sequence apart from an additive disturbance. When this is so, a higher-order scheme is necessary to improve the representation, if an autoregressive scheme indeed exists.

The first- and second-order autoregressive representations are convenient descriptions of how the present relates to the past. Their use often enables the investigator conveniently to summarize the second-order intertemporal dependence of a sequence. Whittle [101, pp. 36–38] and Parzen [76] offer interesting examples.

2.24. Uniqueness of Representation

We have seen that an autoregression may be interpreted as a series of linear filtering operations that turn an autocorrelated sequence X into an uncorrelated sequence ϵ. A question now arises as to whether or not the series of filtering operations that accomplishes this transformation is unique. If not, how many such transformations exist?

To answer this question, we return to the first-order autoregressive representation with spectrum as in Expression (2.27d). Notice that the quantity $(1 - 2\alpha \cos \lambda + \alpha^2)$ is equal to each of the four expressions $|1 - \alpha e^{-i\lambda}|^2$, $|1 - \alpha e^{i\lambda}|^2$, $|\alpha - e^{-i\lambda}|^2$, and $|\alpha - e^{i\lambda}|^2$. The first column of Table 2 lists filters corresponding to these expressions. The second column lists the corresponding autoregressive schemes, and the third column lists the representations of X_t in terms of uncorrelated random variables. Each of these four representations of X has the spectrum given by Expression (2.27d).

Table 2

First-Order Schemes $\qquad |\alpha| < 1$

	$B(\lambda)$	ϵ_t	X_t
1	$1 - \alpha e^{-i\lambda}$	$X_t - \alpha X_{t-1}$	$\sum\limits_{s=0}^{\infty} \alpha^s \epsilon_{t-s}$
2	$1 - \alpha e^{i\lambda}$	$X_t - \alpha X_{t+1}$	$\sum\limits_{s=0}^{\infty} \alpha^s \epsilon_{t+s}$
3	$\alpha - e^{-i\lambda}$	$\alpha X_t - X_{t-1}$	$-\sum\limits_{s=0}^{\infty} \alpha^s \epsilon_{t+s+1}$
4	$\alpha - e^{i\lambda}$	$\alpha X_t - X_{t+1}$	$-\sum\limits_{s=0}^{\infty} \alpha^s \epsilon_{t-s-1}$

For practical purposes we restrict our interest to schemes, as described by Expression (2.26), which omit future values of X. For the first-order scheme, we therefore consider rows 1 and 3 of Table 2. More generally, for an nth-order scheme there are $2n$ solutions to the expression

$$|B(\lambda)|^2 = \left| \sum_{s=0}^{n} b_s e^{-i\lambda s} \right|^2.$$

The moving average in Expression (2.21) is a particularly appealing representation for it permits us to express X_t as the linear combination of present and past disturbances. This representation corresponds to the polynomial solution that has all its roots outside the unit circle. Row 1 satisfies this requirement for the first-order scheme. Row 3 corresponds to a root inside the unit circle, and we note that it leads to an expansion in future values of ϵ_t.

Restricting ourselves to autoregressive schemes in past values of X_t, we note that for a solution with roots outside (inside) the unit circle, X_t is expressible in past (future) values of ϵ_t. A solution with roots both inside and outside the unit circle leads to a representation of X_t in past, present, and future values of ϵ_t. Notice that covariance stationarity precludes any root lying on the unit circle.

In summary, we observe that a particular spectrum may correspond to a number of different stochastic sequences. It is the added restrictions that permit us to identify the sequence with the moving average representation as described in Expression (2.26).

2.25. Trend Elimination and its Effects

The presence of trend in economic time series generally makes it difficult to examine the behavior of the remaining cyclical, seasonal, and irregular components. In addition, if our purpose is to predict the future behavior of the sequence, the mean-square error criterion for choosing an optimal predictor is inapplicable, since it is theoretically unbounded for nonstationary sequences. It is therefore of interest to study methods of reducing nonstationary sequences to stationary ones for which the body of statistical prediction theory applies. We restrict ourselves to a sequence whose mean is a polynomial in t (time).

Suppose that X is a nonstationary sequence such that

$$X_t = \xi_t + \sum_{j=0}^{p} \alpha_j t^j,$$

ξ being a zero mean stochastic sequence, not necessarily covariance stationary, and the pth-order polynomial in t representing a trending mean. Notice that the first difference,

$$\Delta X_t = X_t - X_{t-1} = \xi_t - \xi_{t-1} + \sum_{j=0}^{p} \alpha_j[t^j - (t-1)^j]$$

$$= \Delta\xi_t + \sum_{j=0}^{p} \alpha_j \sum_{k=0}^{j-1} \binom{j}{k+1}(-1)^k t^{j-k-1},$$

is a $(p-1)^{\text{th}}$-order polynomial, implying that first differencing eliminates the term t^p. In a like manner,

$$\Delta^2 X_t = \Delta X_t - \Delta X_{t-1} = X_t - 2X_{t-1} + X_{t-2}$$

eliminates the term in t^{p-1}. In fact, if we perform the differencing operation p times, we have

$$\eta_t = \Delta^p X_t = \sum_{j=0}^{p} \binom{p}{j}(-1)^j X_{t-j} = \Delta^p \xi_t + c,$$

c being a constant determined from the coefficients of the polynomial mean. That is, subjecting the nonstationary sequence X to a linear transformation whose j^{th} coefficient is $\binom{p-1}{j}(-1)^j$, one may successfully eliminate a pth-order polynomial trend in the mean and reduce the residual sequence ξ into a sequence η with mean c. This procedure is called the *variate-difference method*.† Its appeal for economists and statisticians is easily understood since a reasonably simple linear transformation can eliminate a rather complicated trending mean from economic time series. The literature on this method principally concerns how to determine the appropriate order of differencing.

Having removed the trending mean, it remains to discuss the variance of X and, in turn, whether or not η has an autoregressive representation. Suppose that ξ is a covariance stationary uncorrelated sequence. Then $\Delta^p \xi_t$ is a moving average and η is covariance stationary with spectrum

(2.30) $$g_\eta(\lambda) = [2 \sin (\lambda/2)]^{2p} g_\xi(\lambda).$$

Writing $g_\eta(\lambda)$ as

$$g_\eta(\lambda) = (-e^{i\lambda})^p (1 - e^{-i\lambda})^{2p} g_\xi(\lambda),$$

we see that the roots of $(1-z)^{2p}$ lie on the unit circle. Then η cannot have an autoregressive representation.

As another case, suppose that η is an uncorrelated sequence, and take c to be zero for convenience. Then ξ cannot be covariance sta-

---

† For a more complete description of the variate difference method, see Tintner [89].

tionary since g_η is uniform and the Expression (2.30) does not hold for g_ξ unbounded. Indeed one may show that

$$\xi_t = \sum_{j=0}^{p-1} \sum_{s=0}^{\infty} s^j \eta_{t-s}.$$

From this we conclude that the variance of ξ is also a polynomial in t, and in fact that the differencing procedure reduces the order of the polynomial variance. This last fact is true whether or not η is an uncorrelated sequence.

The case of $p = 1$ is of special significance since it has often been used in econometrics. The model

$$\xi_t - \xi_{t-1} = \eta_t,$$

η being an uncorrelated sequence, is called a random walk wherein the variance of ξ increases linearly with time. Note that it is possible to eliminate a linear trend in both mean and variance by one differencing operation. The random walk model has been used by Granger and Morgenstern [31], within the framework of spectral analysis, to study the behavior of stock prices on the New York Stock Exchange.

In general, successive differencing can be used to eliminate polynomial trend in both the mean and variance. The result will generally be a covariance stationary but correlated sequence. In order for η to have an autoregressive representation, it is necessary for its spectrum to be positive, which in turn requires that the order of differencing does not exceed the order of the polynomial in the variance.

It is of interest to study what distortions are induced in ξ as a result of differencing. Suppose that ξ were covariance stationary. Table 3 shows the squared gain. We observe that components with

Table 3

Squared Gain for Variate-Difference Filter

λ	Differences				
	1	2	3	4	5
0	0	0	0	0	0
$\pi/6$	0.268	0.072	0.019	0.005	0.001
$\pi/3$	1.000	1.000	1.000	1.000	1.000
$\pi/2$	2.000	4.000	8.000	16.000	32.000
$2\pi/3$	3.000	9.000	27.000	81.000	243.000
$5\pi/6$	3.732	13.928	51.981	193.995	723.999
π	4.000	16.000	64.000	256.000	1024.000

frequencies below $\pi/3$ are attenuated, whereas those with frequencies greater than $\pi/3$ are amplified. The attenuation and amplification become more pronounced as the number of differences increases. Yule [106] pointed out that if a series contained a periodic component with frequency greater than $\pi/3$, it would be amplified, thus giving the series a more regular periodic appearance. For this reason Tintner [89] cautioned the user of the variate-difference method to remove the seasonal component in order to avoid amplifying its harmonics.

It should also be noted that any periodic component in ξ will be shifted to a higher frequency in η as a result of applying the variate-difference filter. In addition, if ξ were purely random, then η would have a periodic appearance with frequency π, the degree of regularity depending on the order of differencing. These results accord with the slope of the gain always being nonnegative.

Several points of clarification are in order here. To begin with, it is quite reasonable to assume ξ to be nonstationary, for economic time series often exhibit growing deviations around their trend. We have considered polynomial trends here since these are often adequate for explaining growth in mean and variance in economic time series. A further discussion of deviations from stationarity may be found in Whittle [101, pp. 92–94].

As we pointed out at the beginning of this section, reducing a sequence to covariance stationarity permits us to use autoregressive and moving average schemes for prediction purposes. If, for example, we are predicting the pth difference of a covariance stationary sequence, then the mean-square error criterion is applicable since it is finite. Using these predicted differences, one may then derive predictions for the sequence of interest by appropriate substitution. These prediction methods are concisely discussed in Whittle [101], and their application may be found in Box and Jenkins [12] and Couts, Grether, and Nerlove [14].

2.26. Bivariate Spectral Analysis

Economics concerns the study of diverse phenomena whose past, present and future behavior relate in complex ways. The recognition of this multiplicity of relationships among phenomena and among time periods has led to the development of systems of simultaneous linear equations whose functional forms generally follow from economic theory. The ultimate success of this approach resides in the ability to specify the system up to a point where the unexplainable residual behavior of the dependent phenomena are linearly unassociated with any of the phenomena considered and, in addition, are uncorrelated in time.

Contemporary econometrics encompasses two major areas: model specification based on economic theory, and parameter estimation and hypothesis testing using methods of statistical inference based on the theory of multivariate statistical analysis. Were spectral analysis restricted to univariate analysis, its contribution to econometrics would be of marginal value. Determining the extent of autocorrelation and studying the effects of linear filtering permit a more judicious approach to model building and data analysis, but they would not relieve the econometrician's heavy dependence on theory to specify the model.

Multivariate spectral analysis enables the econometrician to gain a reasonably informed idea of what relationships among phenomena the data may imply. It also provides a basis for inferring how phenomena are linearly associated through time. Used in conjunction with economic theory, this knowledge permits the econometrician to eliminate redundancies, to include relevant variables that had formerly not been considered, and generally to understand the relative importance of different phenomena in explaining the different kinds of fluctuations in a particular phenomenon.

As might be expected, multivariate spectral analysis is analogous to multivariate statistical analysis, except that it is applied at each frequency. It is best introduced by studying the linear association between two covariance stationary sequences. Bivariate spectral analysis has been used in oceanography, optics, seismography and communications engineering. Its use in econometrics has been limited to date, but, at the current time, a growing topical literature appears to be emerging. Multivariate spectral analysis beyond the bivariate case will be discussed in theoretical terms in Section 2.30.

Consider two covariance stationary sequences, X and Y, with means

$$E(X_t) = \mu_x, \qquad E(Y_t) = \mu_y,$$

autocovariance functions

$$R_{x,\tau} = E[(X_t - \mu_x)(X_{t+\tau} - \mu_x)], \qquad R_{y,\tau} = E[(Y_t - \mu_y)(Y_{t+\tau} - \mu_y)],$$

autocorrelation functions

$$\rho_{x,\tau} = R_{x,\tau}/R_{x,0}, \qquad \rho_{y,\tau} = R_{y,\tau}/R_{y,0},$$

and spectra

$$g_x(\lambda) = (2\pi)^{-1} \sum_{\tau=-\infty}^{\infty} R_{x,\tau} e^{-i\lambda\tau}, \qquad g_y(\lambda) = (2\pi)^{-1} \sum_{\tau=-\infty}^{\infty} R_{y,\tau} e^{-i\lambda\tau}.$$

In addition, the bivariate sequence has a *covariance function*,

$$R_{xy,\tau} = E[(X_t - \mu_x)(Y_{t+\tau} - \mu_y)],$$

and a *crosscorrelation function*,

$$\rho_{xy,\tau} = R_{xy,\tau}/(R_{x,0}R_{y,0})^{\frac{1}{2}} \qquad |\rho_{xy,\tau}| \leq 1.$$

The crosscorrelation function measures the degree of linear association or correlation between the two sequences for different time lags. It is generally not symmetric around the origin, because the effect of the random variable $X(t)$ on $Y(t + \tau)$ differs from that of the random variable $Y(t)$ on $X(t + \tau)$. The value of τ for which ρ_{xy} has a maximum identifies the time lag with the highest degree of correlation. If this time lag is positive, then the net direction of causality, if it exists, is from X to Y. If it is negative, net causality operates in the opposite direction. The result is predicated on the supposition that the stimulus always occurs prior to the response.

To derive the spectral representation of the covariance function R_{xy}, we use the spectral representation of the random variables X_t and $Y_{t+\tau}$ which lead to

$$
\begin{aligned}
R_{xy,\tau} = \int_0^\pi \int_0^\pi \; & E[\cos \lambda t \, \cos \omega(t + \tau) \, dU_x(\lambda) \, dU_y(\omega) \\
& + \sin \lambda t \, \sin \omega(t + \tau) \, dV_x(\lambda) \, dV_y(\omega) \\
& + \cos \lambda t \, \sin \omega(t + \tau) \, dU_x(\lambda) \, dV_y(\omega) \\
& + \sin \lambda t \, \cos \omega(t + \tau) \, dV_x(\lambda) \, dU_y(\omega)].
\end{aligned}
$$

For this expression to be independent of t, we require the covariance relationships

$$E[dU_x(\lambda) \, dU_y(\omega)] = E[dV_x(\lambda) \, dV_y(\omega)] = \begin{cases} dC(\lambda) & \lambda = \omega \\ 0 & \lambda \neq \omega, \end{cases}$$

and

$$E[dU_x(\lambda) \, dV_y(\omega)] = -E[dV_x(\lambda) \, dU_y(\omega)] = \begin{cases} dQ(\lambda) & \lambda = \omega \\ 0 & \lambda \neq \omega, \end{cases}$$

so that

$$R_{xy,\tau} = \int_0^\pi [\cos \lambda\tau \, dC(\lambda) + \sin \lambda\tau \, dQ(\lambda)].$$

The absence of regular periodic components implies that the functions C and Q are continuous, so that

$$dC(\lambda) = c(\lambda)\, d\lambda, \qquad dQ(\lambda) = q(\lambda)\, d\lambda.$$

The functions c and q are called the co-spectrum and the quadrature-spectrum, respectively. Defining

$$g_{xy}(\lambda) = c(\lambda) - iq(\lambda),$$

$$c(\lambda) = c(-\lambda), \qquad q(\lambda) = -q(-\lambda) \qquad -\pi \le \lambda \le \pi,$$

we have

$$R_{xy,\tau} = \int_{-\pi}^{\pi} g_{xy}(\lambda)e^{i\lambda\tau}\, d\lambda,$$

where g_{xy} is called the cross spectrum. As before, we have the inverse relationship

$$g_{xy}(\lambda) = (2\pi)^{-1} \sum_{\tau=-\infty}^{\infty} R_{xy,\tau}e^{-i\lambda\tau}.$$

To formalize the linear relationship between the sequences X and Y into parametric representation, we may write

(2.31)
$$Y_t = \sum_{s=-\infty}^{\infty} a_s X_{t-s} + \eta_t,$$

$$\sum_{s=-\infty}^{\infty} a_s^2 < \infty,$$

where the covariance stationary sequence $\{\eta_t\}$ has a zero mean and is uncorrelated with X. The sequence $\{\eta_t\}$ accounts for the nonlinear relationship between X and Y as well as for the influence of other processes on the behavior of Y. If Y were simply a transformation of X produced by linear filtering, then the sequence $\{\eta_t\}$ would vanish. This was the case in Section 2.17.

The covariance function may now be represented as

(2.32)
$$R_{xy,\tau} = \sum_{s=-\infty}^{\infty} a_s R_{x,\tau-s},$$

so that

(2.33)
$$g_{xy}(\lambda) = A(\lambda)g_x(\lambda),$$

where the frequency response function is

$$A(\lambda) = \sum_{s=-\infty}^{\infty} a_s e^{-i\lambda s}.$$

As mentioned earlier, the frequency response function measures attenuation and delay via its gain G and phase angle ϕ, respectively, where

(2.34) $$A(\lambda) = G(\lambda)e^{i\phi(\lambda)}.$$

Using Expression (2.33), these measures are defined as

(2.35a) $$G(\lambda) = [c^2(\lambda) + q^2(\lambda)]^{\frac{1}{2}}/g_x(\lambda),$$

(2.35b) $$\phi(\lambda) = \tan^{-1}[-q(\lambda)/c(\lambda)].$$

The significance of the frequency response function depends on the extent of linear association between the harmonic components in X and their corresponding components in Y. Consider the correlation between the components

(2.36a) $$X_t(\lambda) = \cos \lambda t \, dU_x(\lambda) + \sin \lambda t \, dV_x(\lambda)$$

and

(2.36b) $$Y_t(\lambda) = \cos [\lambda t + \phi(\lambda)] \, dU_y(\lambda) + \sin [\lambda t + \phi(\lambda)] \, dV_y(\lambda)$$

in X and Y respectively. Using the above definitions of spectra, we have†

$$\text{corr } [X_t(\lambda), Y_t(\lambda)] = [c(\lambda) \cos \phi(\lambda) + q(\lambda) \sin \phi(\lambda)]/[g_x(\lambda)g_y(\lambda)]^{\frac{1}{2}}.$$

The correlation is maximized when $\phi(\lambda)$ is defined as in Expression (2.35b), so that

(2.37) $$\gamma(\lambda) = \max_{\phi(\lambda)} \{\text{corr } [X_t(\lambda), Y_t(\lambda)]\}$$

$$= \{[c^2(\lambda) + q^2(\lambda)]/[g_x(\lambda)g_y(\lambda)]\}^{\frac{1}{2}},$$

which is called the coherence at frequency λ. This result implies that the correlation between $X_t(\lambda)$ and $Y_t(\lambda)$ is maximized by treating the sinusoid in Y with frequency λ as having a phase difference $\phi(\lambda)$ as given by Expression (2.35b). To determine the significance of a lag appropriately, it is therefore necessary to consider the complex quantity $\gamma(\lambda)e^{-i\phi(\lambda)}$.

Three particular shapes of the phase angle curve are worth emphasizing. If this curve has a constant slope, then the time delay is uniform at each frequency. If the slope increases with increasing frequencies, then high-frequency components are delayed longer (in time) than low-frequency ones. If the slope decreases with increasing frequencies, then low-frequency components are delayed longer (in time) than high-frequency ones.

---

† See Hannan [41, p. 67].

Using Expression (2.35a), we may write

(2.38) $$G(\lambda) = \gamma(\lambda)[g_y(\lambda)/g_x(\lambda)]^{\frac{1}{2}}.$$

Notice that although the coherence may be constant over the entire frequency range, the corresponding gain need not necessarily be so. This means that the same linear association exists at all frequencies, but that some frequency components may be amplified or attenuated more than others.

Figure 14 shows the sample coherence, gain, and phase angle for shipments and inventories. Notice the high sample coherence at low frequencies, indicating a strong linear association. More interesting is the unusually low sample coherence at the six-month frequency $\pi/3$. This suggests that the six-month cycle in shipments is not highly correlated with the six-month cycle in inventories.

The sample gain indicates that low frequencies ($\leq \pi/48$) are amplified and high frequencies are attenuated. This means that long-term increases in shipments are associated with even larger increases in inventories, whereas short-term increases in shipments are associated with relatively meager increases in inventories. Notice also that the sample phase angle curve shows several large phase shifts.

Returning to Expression (2.30), one may show that

(2.39a) $$R_{y,\tau} = \sum_{s_1=-\infty}^{\infty} a_{s_1} \sum_{s_2=-\infty}^{\infty} a_{s_2} R_{x,\tau+s_1-s_2} + R_{\eta,\tau},$$

where R_η is the autocovariance function of the sequence $\{\eta_t\}$. Taking Fourier transforms yields

(2.39b) $$g_y(\lambda) = |A(\lambda)|^2 g_x(\lambda) + g_\eta(\lambda),$$

where g_η is the spectrum of the sequence $\{\eta_t\}$. Combining Expressions (2.34), (2.35a), (2.37), and (2.40) gives us

$$g_\eta(\lambda) = g_y(\lambda)[1 - \gamma^2(\lambda)].$$

Notice that a unit coherence implies that the spectrum of η vanishes. A perfect correlation then exists at each frequency in X and Y. It is interesting to observe that the covariance function in Expression (2.32) does not reflect this unit correlation. Indeed it may assume arbitrary values including zero for some values of τ, while the unit coherence still prevails in the frequency domain.

If X yields a linear representation of Y up to an uncorrelated disturbance η, then the spectrum g_η is uniform over the frequency

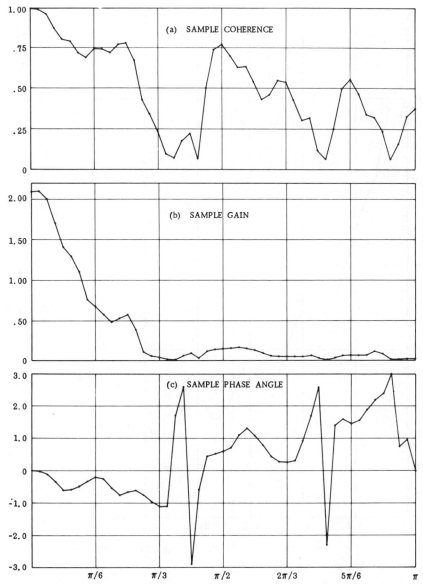

Fig. 14 Manufacturers' shipments and inventories of durable goods, seasonally unadjusted

range $(-\pi, \pi)$. This result is very significant because it gives us a way
of determining how adequately X accounts for the structure in Y.
If η is not uncorrelated, then some autocorrelated structure remains in
Y after the effects of X have been removed. It is then necessary to
introduce other processes to make the residuals uncorrelated.

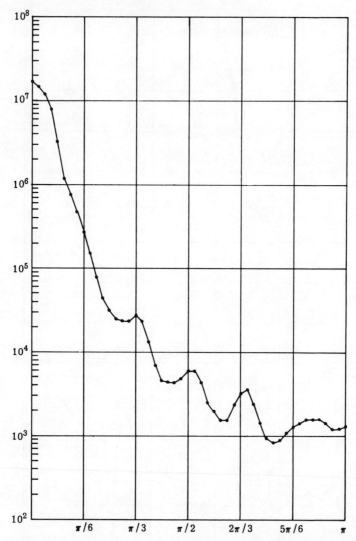

Fig. 15 Sample residual spectrum of manufacturers' inventories of durable goods, seasonally unadjusted

Figure 15 shows the sample residual spectrum for inventories. The magnitude varies from less than 10^3 to over 10^7, indicating that supplemental phenomena are needed to explain the structure of the inventory residuals.

One further point of clarification is worthwhile here. Even if the residual spectrum is uniform, additional phenomena may play a role in explaining variation in Y. Suppose that the correct representation is

$$Y_t = \sum_{s=-\infty}^{\infty} a_s X_{t-s} + \sum_{s=-\infty}^{\infty} b_s Z_{t-s} + \epsilon_t.$$

Assuming for simplicity that X and Z are mutually uncorrelated sequences, we have

$$g_\eta(\lambda) = g_y(\lambda)[1 - \gamma_{xy}^2(\lambda)] = \gamma_{zy}^2(\lambda)g_y(\lambda) + g_\epsilon(\lambda).$$

If g_η is flat, then X accounts for all the autocorrelation structure in Y, but including Z reduces the variation in Y even more.

2.27. The Effects of Seasonal Adjustment

Many economic time series are available both seasonally un-adjusted and adjusted. The adjusted series permits one to examine variations unimpaired by seasonal movements. It is of interest to ask what modifications are made in the frequency content of the unadjusted series in addition to removing the seasonal components. Nerlove [67] has studied this question in some detail using spectral methods. Here we examine this same question in a similar way.

We consider the unadjusted series as the input to a linear filter and the adjusted series as the output. The Bureau of the Census and the Bureau of Labor Statistics actually use *nonlinear* seasonal adjust-ment procedures; thus, we should regard the linear filtering representa-tion as only an approximation. There are certain features of a filtering procedure we deem necessary if the components of the series other than the seasonal one are to be preserved in the adjusted series. We expect the coherence at all frequencies but the seasonals to be unity. This insures the preservation of the linear association between input and output. The gain should have a similar appearance, thereby pre-serving the magnitude of the variation at frequencies other than the seasonals. The phase angle should be zero, so as to avoid any shifting.

Figure 16 shows the sample coherence, gain, and phase angle for shipments, seasonally unadjusted and adjusted. The sample coherence, while not unity, is high at all but the seasonals. The sample gain is also high at all but the seasonals, but clearly attenuation has taken place. Hence we expect the adjusted series to be somewhat smoother than the unadjusted one. Notice that the only phase shifts occur at the seasonals. From studying these curves we may conclude that linear association is preserved fairly well, some attenuation occurs in the magnitude of variation, and no important phase shifts occur.

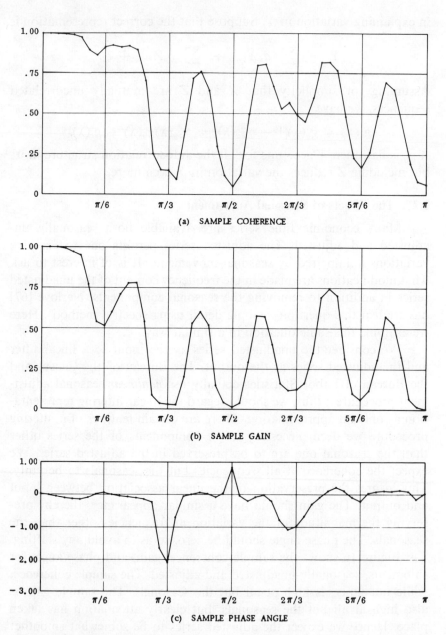

(a) SAMPLE COHERENCE

(b) SAMPLE GAIN

(c) SAMPLE PHASE ANGLE

Fig. 16 Manufacturers' shipments of durable goods, seasonally unadjusted and adjusted

2.28. The Effect of Linear Filtering on the Relationship between X and Y

We now consider the relationship between the sequences $\tilde{X}$ and $\tilde{Y}$ where

$$\tilde{X}_t = \sum_{s=p_1}^{q_1} a_s X_{t-s}, \qquad \tilde{Y}_t = \sum_{s=p_2}^{q_2} b_s Y_{t-s}.$$

This leads to

$$\tilde{R}_{x,\tau} = \sum_{s_1,s_2=-p_1}^{q_1} a_{s_1} a_{s_2} R_{x,\tau+s_1-s_2},$$

$$\tilde{R}_{y,\tau} = \sum_{s_1,s_2=-p_2}^{q_2} b_{s_1} b_{s_2} R_{y,\tau+s_1-s_2},$$

$$\tilde{R}_{xy,\tau} = \sum_{s_1=-p_1}^{q_1} \sum_{s_2=-p_2}^{q_2} a_{s_1} b_{s_2} R_{xy,\tau+s_1-s_2},$$

$$\tilde{g}_x(\lambda) = |A(\lambda)|^2 g_x(\lambda),$$

$$\tilde{g}_y(\lambda) = |B(\lambda)|^2 g_y(\lambda),$$

$$\tilde{g}_{xy}(\lambda) = A^*(\lambda)B(\lambda)g_{xy}(\lambda),$$

so that

$$\tilde{G}_{xy}(\lambda) = |A(\lambda)|^{-1}|B(\lambda)|G_{xy}(\lambda),$$

$$\tan \tilde{\phi}(\lambda) = [A_1(\lambda)B_2(\lambda) + A_2(\lambda)B_1(\lambda)]/[A_1(\lambda)B_1(\lambda) + A_2(\lambda)B_2(\lambda)],$$

$$\tilde{\gamma}_{xy}(\lambda) = \gamma_{xy}(\lambda),$$

where

$$A(\lambda) = A_1(\lambda) + iA_2(\lambda) = \sum_{s=-p_1}^{q_1} a_s e^{-i\lambda s},$$

$$B(\lambda) = B_1(\lambda) + iB_2(\lambda) = \sum_{s=-p_2}^{q_2} b_s e^{-i\lambda s}.$$

Notice that the coherence is preserved under linear transformation. Furthermore, if both X and Y are subjected to the same transformation so that $A(\lambda) = B(\lambda)$, the gain and phase angle are also preserved. In addition, the distributed lag model becomes

$$(2.40) \qquad \tilde{Y}_t = \sum_{s=-\infty}^{\infty} a_s \tilde{X}_{t-s} + \tilde{\eta}_t.$$

The role of linear filtering is of special significance when X and Y are nonstationary. By a judicious choice of filter, one can often identically transform the time series on X and Y to covariance stationarity and use Expression (2.40) as the distributed lag scheme. This is particularly convenient when we wish to estimate the coefficients in Expression (2.31), for we see that the corresponding coefficients are identical in both Expressions (2.31) and (2.40). Chapter 4 describes the estimation techniques in detail and makes use of this desirable property.

As Section 2.26 has shown, the variate-difference method can eliminate polynomial trend. This approach has disadvantages, however, when estimating spectra and so a quasi-differencing method seems more appropriate. Because of its close connection with spectrum estimation, we defer the description of quasi-differencing to Section 3.10.

2.29. Econometric Models and Spectral Analysis

In general, it is naive to suppose that the relationships between economic phenomena can be described by single equations. Contemporary econometrics stresses the need to relate phenomena through systems of simultaneous equations in which causality is permitted to operate in both directions. For example, A may be an endogenous variable that reacts to changes in the exogenous variable B, whereas B may be an endogenous variable that changes in response to past values of A. Systems of simultaneous equations permit us to account for the presence of this *feedback* among phenomena. Demand and supply curves are typical examples; the student may wonder, upon examining them, whether price influences quantity or quantity influences price. Each influences the other, of course, and thus they make up a feedback loop.

It is natural to ask how spectral methods can improve our understanding of the superimposed effects of feedback. To study this question, we consider the following description of macro-economic behavior.†
Let X, Y, U, and V denote income, consumption, autonomous investment, and residual consumption respectively. Assume them to be related by

(2.41a) $$X_t = Y_t + a(X_{t-1} - X_{t-2}) + U_t,$$

(2.41b) $$Y_t = bX_{t-1} + V_t, \qquad 0 < a, b < 1,$$

where $a(X_{t-1} - X_{t-2})$ denotes induced investment in period t. The

---

† See Cox and Miller [15] for an example concerning the price-quantity relationship.

uncorrelated sequences U and V with variances σ_u^2 and σ_v^2, respectively, are mutually uncorrelated as are U with Y, and V with X.

Rearranging terms yields the difference equations

(2.42a)
$$X_t - (a + b)X_{t-1} + aX_{t-2} = U_t + V_t,$$

(2.42b)
$$Y_t - (a + b)Y_{t-1} + aY_{t-2} = bU_{t-1} + V_t - a(V_{t-1} - V_{t-2}).$$

Notice that the difference equation in X is an autoregressive representation. In order for X and Y to be covariance stationary, we require the roots of the polynomial

$$1 - (a + b)\xi + a\xi^2 = 0$$

to lie outside the unit circle.

The spectral representations are

$$h(\lambda)z_x(\lambda) = z_u(\lambda) + z_v(\lambda),$$
$$h(\lambda)z_y(\lambda) = be^{-i\lambda}z_u(\lambda) + [1 - a(e^{-i\lambda} - e^{-i2\lambda})]z_v(\lambda),$$
$$h(\lambda) = 1 - (a + b)e^{-i\lambda} + ae^{-i2\lambda},$$

which lead to the spectra

$$g_x(\lambda) = (\sigma_u^2 + \sigma_v^2)k(\lambda),$$
$$g_y(\lambda) = (b^2\sigma_u^2 + |1 - a(e^{-i\lambda} - e^{-i2\lambda})|^2\sigma_v^2)k(\lambda),$$
$$g_{xy}(\lambda) = \{be^{-i\lambda}\sigma_u^2 + [1 - a(e^{-i\lambda} - e^{-i2\lambda})]\sigma_v^2\}k(\lambda),$$
$$k(\lambda) = 1/[2\pi|h(\lambda)|^2].$$

The gain, phase angle, and coherence are

$$G(\lambda) = |\sigma_v^2 + (b\sigma_u^2 - a\sigma_v^2)e^{-i\lambda} + a\sigma_v^2 e^{-i2\lambda}|/(\sigma_u^2 + \sigma_v^2),$$

$$\tan\phi(\lambda) = [(b\sigma_u^2 - a\sigma_v^2)\sin\lambda + a\sigma_v^2\sin 2\lambda]$$
$$/[\sigma_v^2 + (b\sigma_u^2 - a\sigma_v^2)\cos\lambda + a\sigma_v^2\cos 2\lambda],$$

$$\gamma(\lambda) = G(\lambda)[(\sigma_u^2 + \sigma_v^2)/(b^2\sigma_u^2 + |1 - a(e^{-i\lambda} - e^{-i2\lambda})|^2\sigma_v^2)]^{\frac{1}{2}}.$$

Notice that if $a/b = \sigma_u^2/\sigma_v^2$ then the gain has a minimum at $\lambda = \pi/2$, implying that low- and high-frequency fluctuations are less attenuated than mid-band frequency fluctuations.

2.30. Multivariate Spectral Analysis†

It is often the case in economics that relying on bivariate models to explain behavior leads to erroneous results. Suppose that the sequences X and Y are highly correlated with a third sequence X'. The coherence function between X and Y will consequently exhibit a high degree of association at all frequencies for which the corresponding components in X and X' are highly coherent. This occurs because of the failure to remove the association between X and X' at these frequencies. A common example is that the current value of X, namely X_t, is to a major extent a consequence of past values of X', namely $X'_{t-1}, X'_{t-2}, \ldots$. It may alternatively be the case that X and X' show only a weak association with Y separately but, when considered jointly, suffice to explain a considerable amount of the behavior of Y.

In spectral analysis, the problem of association has an added dimension—frequency. One may easily imagine a problem in which X and Y have high coherence at low frequencies and low coherence at high frequencies, whereas the association between X' and Y is quite the opposite. For example, the phenomena that influence long-term fluctuations in economic time series are probably different from the phenomena that influence short-term fluctuations.

In order for us to develop the multivariate generalization of spectral analysis, it is appropriate to consider the reduced form of a set of simultaneous distributed lag equations. Since our initial purpose is to derive the multiple and partial coherence functions for each equation, it is more convenient to begin with one such equation. Later we shall consider a set of simultaneous equations and develop several points relating to them in the frequency domain.

Consider

$$(2.43) \qquad Y_t = \sum_{s=-r_1}^{r_2} \mathbf{a}_s \mathbf{X}_{t-s} + \eta_t,$$

where $\mathbf{X}_t$ is an $N \times 1$ vector of exogenous variables

$$\mathbf{X}_t = \begin{bmatrix} X_{1,t} \\ \vdots \\ X_{p,t} \\ \hline X_{p+1,t} \\ \vdots \\ X_{N,t} \end{bmatrix} = \begin{bmatrix} \mathbf{X}_{1,t} \\ \hline \mathbf{X}_{2,t} \end{bmatrix}$$

† See Enochson [22] and Goodman [28].

with zero mean vector; Y_t and η_t are scalars with zero means; and $\mathbf{a}_s$ is a $1 \times N$ vector of coefficients

$$\mathbf{a}_s = [a_{1,s} \ldots a_{p,s} \mid a_{p+1,s} \ldots a_{N,s}].$$

The partitioning of the vectors simplifies the following analysis. We also specify that

(2.44)
$$E(X_{j,t}\eta_{t+\tau}) = 0 \qquad j = 1, 2, \ldots, N; \; \tau = 0, \pm 1, \pm 2, \ldots, \pm \infty.$$

If $\{\eta_t\}$ is an uncorrelated sequence, then we may regard the vector sequence $\mathbf{X}$ as being a set of predetermined variables, thus permitting past values of Y to enter. If $\{\eta_t\}$ is not uncorrelated, then past values of the dependent variable are not permitted since they conflict with Expression (2.44).

We define the covariance matrices

$$\mathbf{R}_{xx,\tau} = E(\mathbf{X}_t \mathbf{X}'_{t+\tau})$$

$$= \begin{bmatrix} R_{11,\tau} & \cdots & R_{1p,\tau} & R_{1,p+1,\tau} & \cdots & R_{1N,\tau} \\ \vdots & & \vdots & \vdots & & \vdots \\ R_{p1,\tau} & \cdots & R_{pp,\tau} & R_{p,p+1,\tau} & \cdots & R_{pN,\tau} \\ \hline R_{p+1,1,\tau} & \cdots & R_{p+1,p,\tau} & R_{p+1,p+1,\tau} & \cdots & R_{p+1,N,\tau} \\ \vdots & & \vdots & \vdots & & \vdots \\ R_{N1,\tau} & \cdots & R_{Np,\tau} & R_{N,p+1,\tau} & \cdots & R_{NN,\tau} \end{bmatrix},$$

$$\mathbf{R}_{xy,\tau} = E(\mathbf{X}_t Y_{t+\tau}) = \begin{bmatrix} R_{1y,\tau} \\ \vdots \\ R_{py,\tau} \\ \hline R_{p+1,y,\tau} \\ \vdots \\ R_{Ny,\tau} \end{bmatrix},$$

$$R_{jk,\tau} = R_{x_j x_k,\tau}, \qquad R_{jy,\tau} = R_{x_j y,\tau},$$

so that

$$R_{y,\tau} = \sum_{\nu,s=-r_1}^{r_2} \mathbf{a}_s \mathbf{R}_{xx,\tau+s-\nu} \mathbf{a}'_\nu + R_{\eta,\tau},$$

$$\mathbf{R}_{xy,\tau} = \sum_{s=-r_1}^{r_2} \mathbf{R}_{xx,\tau-s} \mathbf{a}'_s.$$

The corresponding spectral relationships are

$$g_y(\lambda) = \mathbf{A}^*(\lambda)\mathbf{g}_{xx}(\lambda)\mathbf{A}(\lambda) + g_\eta(\lambda), \qquad \mathbf{g}_{xy}(\lambda) = \mathbf{g}_{xx}(\lambda)\mathbf{A}(\lambda),$$

$$\mathbf{g}_{xx}(\lambda) = \begin{bmatrix} g_{11}(\lambda) & \cdots & g_{1p}(\lambda) & g_{1,p+1}(\lambda) & \cdots & g_{1N}(\lambda) \\ \vdots & & \vdots & \vdots & & \vdots \\ g_{p1}(\lambda) & \cdots & g_{pp}(\lambda) & g_{p,p+1}(\lambda) & \cdots & g_{pN}(\lambda) \\ \hline g_{p+1,1}(\lambda) & \cdots & g_{p+1,p}(\lambda) & g_{p+1,p+1}(\lambda) & \cdots & g_{p+1,N}(\lambda) \\ \vdots & & \vdots & \vdots & & \vdots \\ g_{N1}(\lambda) & \cdots & g_{Np}(\lambda) & g_{N,p+1}(\lambda) & \cdots & g_{NN}(\lambda) \end{bmatrix},$$

$$= \begin{bmatrix} \mathbf{h}_{11}(\lambda) & \mathbf{h}_{12}(\lambda) \\ \hline \mathbf{h}_{21}(\lambda) & \mathbf{h}_{22}(\lambda) \end{bmatrix},$$

$$\mathbf{g}_{xy}(\lambda) = \begin{bmatrix} g_{1y}(\lambda) \\ \vdots \\ g_{py}(\lambda) \\ \hline g_{p+1,y}(\lambda) \\ \vdots \\ g_{Ny}(\lambda) \end{bmatrix} = \begin{bmatrix} \mathbf{m}_1(\lambda) \\ \hline \mathbf{m}_2(\lambda) \end{bmatrix},$$

$$g_{jk}(\lambda) = g_{x_j x_k}(\lambda), \qquad g_{jy}(\lambda) = g_{x_j y}(\lambda),$$

$$\mathbf{A}(\lambda) = \sum_{s=-r_1}^{r_2} \mathbf{a}_s' e^{-i\lambda s},$$

$$\mathbf{A}^*(\lambda) = \sum_{s=-r_1}^{r_2} \mathbf{a}_s e^{i\lambda s}.$$

We require the matrix $\mathbf{g}_{xx}(\lambda)$ to be nonsingular which, in turn, requires $\mathbf{R}_{xx}$ to be positive definite. Notice that $\mathbf{A}(\lambda)$ is an $N \times 1$ column vector whereas $\mathbf{A}^*(\lambda)$ is a $1 \times N$ row vector. If the N exogenous sequences are mutually uncorrelated, then

$$E(X_{j,t}X_{k,s}) = \begin{cases} R_{jj,t-s} & j = k \\ 0 & j \neq k \end{cases} \qquad t, s = 0, \pm 1, \pm 2, \ldots, \pm \infty,$$

so that $\mathbf{R}_{xx,\tau}$ and $\mathbf{g}_{xx}(\lambda)$ are diagonal matrices.

The first question of interest is, to what extent do the $N(r_1 + r_2 + 1)$ exogenous variables explain the behavior of Y_t? The strength of linear association is measured by the *multiple coherence function.* Defining

$$Z_t = \sum_{s=-r_1}^{r_2} a_s X_{t-s},$$

we have

$$Y_t = Z_t + \eta_t,$$
$$g_z(\lambda) = \mathbf{A}^*(\lambda)\mathbf{g}_{xx}(\lambda)\mathbf{A}(\lambda),$$
$$g_{zy}(\lambda) = \mathbf{A}^*(\lambda)\mathbf{g}_{xy}(\lambda) = \mathbf{A}^*(\lambda)\mathbf{g}_{xx}(\lambda)\mathbf{A}(\lambda),$$

so that the multiple coherence is

(2.45)
$$\gamma_{y\cdot1,2,\ldots,N}(\lambda) = |g_{zy}(\lambda)|/[g_y(\lambda)g_z(\lambda)]^{\frac{1}{2}}$$
$$= \{[\mathbf{g}_{xy}^*(\lambda)\mathbf{g}_{xx}^{-1}(\lambda)\mathbf{g}_{xy}(\lambda)]/g_y(\lambda)\}^{\frac{1}{2}},$$

$\mathbf{g}_{xy}^*(\lambda)$ being the complex conjugate of the transpose of $\mathbf{g}_{xy}(\lambda)$.

It is often of interest to define the extent to which some subset of the N exogenous phenomena suffice to explain the behavior of Y_t. To investigate this problem, we examine the *marginal multiple coherence,* which is derived as follows. Suppose we are interested in the explanatory power of the first p sequences. Then we need only consider the submatrices $\mathbf{h}_{11}(\lambda)$ and $\mathbf{m}_1(\lambda)$ so that the marginal multiple coherence is

(2.46)
$$\gamma_{y\cdot1,2,\ldots,p}(\lambda) = \{[\mathbf{m}_1^*(\lambda)\mathbf{h}_{11}^{-1}(\lambda)\mathbf{m}_1(\lambda)]/g_y(\lambda)\}^{\frac{1}{2}}.$$

As mentioned earlier, the true strength of association between phenomena is often obscured by the common association of endogenous and exogenous variables with other phenomena. Suppose we wish to establish the degree of linear association between Y and $\mathbf{X}_2$, but free of the influence of $\mathbf{X}_1$. We first consider the vector

(2.47)
$$\mathbf{X}_{2,t} = \sum_{s=-q_1}^{q_2} \mathbf{b}_s \mathbf{X}_{1,t-s} + \boldsymbol{\xi}_t,$$

where

$$\mathbf{b}_s = \begin{bmatrix} b_{p+1,1,s} & b_{p+1,2,s} & \cdots & b_{p+1,p,s} \\ b_{p+2,1,s} & b_{p+2,2,s} & & \vdots \\ \vdots & & \ddots & \\ b_{N\,1,s} & \cdots & & b_{Np,s} \end{bmatrix},$$

$$\boldsymbol{\xi}_t = \begin{bmatrix} \xi_{p+1,t} \\ \xi_{p+2,t} \\ \vdots \\ \xi_{N,t} \end{bmatrix},$$

$$E(\boldsymbol{\xi}_t) = 0, \qquad E(\mathbf{X}_{1,t}\boldsymbol{\xi}_s) = 0 \qquad s, t = 0, \pm1, \pm2, \ldots, \pm\infty.$$

This leads to

$$\mathbf{h}_{12}(\lambda) = \mathbf{h}_{11}(\lambda)\mathbf{B}(\lambda),$$

$$\mathbf{h}_{21}(\lambda) = \mathbf{B}^*(\lambda)\mathbf{h}_{11}(\lambda),$$

$$\mathbf{h}_{22}(\lambda) = \mathbf{B}^*(\lambda)\mathbf{h}_{11}(\lambda)\mathbf{B}(\lambda) + \mathbf{g}_{\xi\xi}(\lambda),$$

$$\mathbf{B}(\lambda) = \sum_{s=-q_1}^{q_2} \mathbf{b}_s' e^{-i\lambda s},$$

$$\mathbf{B}^*(\lambda) = \sum_{s=-q_1}^{q_2} \mathbf{b}_s e^{i\lambda s},$$

so that

(2.48) $$\mathbf{g}_{\xi\xi}(\lambda) = \mathbf{h}_{22}(\lambda) - \mathbf{h}_{21}(\lambda)\mathbf{h}_{11}^{-1}(\lambda)\mathbf{h}_{12}(\lambda).$$

In addition we have

$$\mathbf{g}_{\xi y}(\lambda) = \mathbf{m}_2(\lambda) - \mathbf{h}_{21}(\lambda)\mathbf{h}_{11}^{-1}(\lambda)\mathbf{m}_1(\lambda).$$

Now the set of sequences ξ contains that part of the set of sequences $\mathbf{X}_2$ not linearly associated with $\mathbf{X}_1$.

We next investigate the relationship

$$Y_t = \sum_{s=-l_1}^{l_2} \mathbf{c}_s \xi_{t-s} + \varphi_t,$$

where

$$\mathbf{c}_s = \begin{bmatrix} c_{p+1,s} \\ c_{p+2,s} \\ \vdots \\ c_{N,s} \end{bmatrix},$$

$$E(\xi_t \varphi_s) = 0 \qquad t, s = 0, \pm 1, \pm 2, \ldots, \pm\infty.$$

Then, as in the case of multiple coherence, one may show that the *partial coherence* is

$$\gamma_{y \cdot p+1, p+2, \ldots, N \mid 1, 2, \ldots, p}(\lambda) = \{[\mathbf{g}_{\xi y}^*(\lambda)\mathbf{g}_{\xi\xi}^{-1}(\lambda)\mathbf{g}_{\xi y}(\lambda)]/g_y(\lambda)\}^{\frac{1}{2}},$$

the matrix $\mathbf{g}_{\xi y}^*(\lambda)$ denoting the complex conjugate of the transpose of $\mathbf{g}_{\xi y}(\lambda)$. A comparison of the multiple and partial coherence functions reveals which frequency bands are affected by removing the influence of $\mathbf{X}_1$.

The foregoing described the associative aspects of multivariate analysis in a single reduced form equation. An econometric model generally contains a set of simultaneous equations based on structural relationships among economic phenomena. It is of interest to study how well a model specifying certain exogenous phenomena accounts for the autocorrelation structure of the specified endogenous phenomena. An ancillary question of interest concerns the relationship among the equations of the model.

Consider the set of p equations,

$$(2.49) \qquad \mathbf{Y}_t = \sum_{s=-r_1}^{r_2} \mathbf{a}_s \mathbf{X}_{t-s} + \boldsymbol{\eta}_t,$$

where $\mathbf{Y}_t$ is a $p \times 1$ vector of endogenous variables, $\mathbf{X}_t$ is an $N \times 1$ vector of exogenous variables, $\boldsymbol{\eta}_t$ is a $p \times 1$ vector of residuals such that

$$E(\boldsymbol{\eta}_t) = 0, \qquad E(\mathbf{X}_t \boldsymbol{\eta}_s) = 0, \qquad s, t = 0, \pm 1, \pm 2, \ldots, \pm \infty.$$

The $N \times N$ matrix $\mathbf{a}_s$ is

$$\mathbf{a}_s = \begin{bmatrix} a_{11,s} & a_{12,s} & \cdots & a_{1N,s} \\ a_{21,s} & a_{22,s} & & \vdots \\ \vdots & & \ddots & \\ a_{N1,s} & \cdots & & a_{NN,s} \end{bmatrix}.$$

For convenience we assume there are $r_1 + r_2 + 1$ distributed lags in each equation and for each exogenous variable. Notice that (2.49) is equivalent to (2.47) except for notation.

The residual spectrum matrix of $\boldsymbol{\eta}$ is

$$(2.50) \qquad \mathbf{g}_{\eta\eta}(\lambda) = \mathbf{g}_{yy}(\lambda) - \mathbf{g}_{xy}^{*}(\lambda) \mathbf{g}_{xx}^{-1}(\lambda) \mathbf{g}_{xy}(\lambda),$$

which is derived similarly to (2.48). We assume $\mathbf{g}_{\eta\eta}(\lambda)$ to be nonsingular. This implies that no identities enter our model for then $\mathbf{g}_{\eta\eta}(\lambda)$ would have a column and row of zeros and would therefore be singular.

If $\mathbf{g}_{\eta\eta}(\lambda)$ is uniform for all λ in $(-\pi, \pi)$, then $\{\mathbf{X}_t\}$ completely accounts for the autocorrelation structure of $\{\mathbf{Y}_t\}$. As mentioned in Section 2.26, however, the inclusion of supplemental phenomena may reduce the residual variation.

If $\mathbf{g}_{\eta\eta}(\lambda)$ is a diagonal matrix for all λ in $(-\pi, \pi)$, then we may apply unweighted linear least-squares methods to the individual equations to derive efficient coefficient estimates. Other situations may also occur. For example, some but not all of the equations may be uncorrelated. Identifying and removing the uncorrelated ones reduces the dimensionality of the estimation problem for the remaining equations.

These interpretations of the residual spectrum matrix $\mathbf{g}_{\eta\eta}(\lambda)$ make its estimation and testing desirable. Section 3.21 describes the estimation procedure which is analogous to the bivariate case. Testing procedures are described in Sections 3.22 and 4.13.

3 Spectrum Analysis

3.1. Introduction

This chapter describes *spectrum analysis*—the problems of statistical inference as they apply to the spectrum and cross spectrum. To cover this discipline adequately, it is necessary to precede the presentation with a fairly detailed description of *correlation analysis*—the problems of statistical inference as they apply to the autocovariance and auto-correlation functions. By proceeding in this way, we shall find that the salient features of each analysis become more apparent and the value of each can be assessed more easily.

Our exposition will be simplified by using the moving average representation

$$(3.1) \qquad X_t = \sum_{s=0}^{\infty} a_s \epsilon_{t-s} \qquad t = 0, \pm 1, \pm 2, \ldots, \pm \infty,$$

with the elements of $\{\epsilon_t\}$ being mutually uncorrelated and stationary to the fourth order so that

$$(3.2) \qquad E(\epsilon_t) = 0; \qquad E(\epsilon_t \epsilon_{t+s}) = \begin{cases} 1, & s = 0 \\ 0, & s \neq 0 \end{cases},$$

$$E(\epsilon_t^4) = \varkappa + 3 < \infty,$$

$$R_\tau = \sum_{s=0}^{\infty} a_s a_{s+\tau},$$

$$g(\lambda) = (2\pi)^{-1} \left| \sum_{s=0}^{\infty} a_s e^{-i\lambda s} \right|^2.$$

Notice that X is not necessarily a discrete linear process and consequently is not necessarily strictly stationary as was shown in Section 2.20. We require that its first four moments be time invariant; then X may be evolutionary, albeit in its highest-order moments. This representation of X is often used in economics and is a convenient model for studying the statistical properties of correlation and spectrum analysis. It should be mentioned that if $\{\epsilon_t\}$ is Gaussian white noise, then $\varkappa$ is zero and X is a discrete linear process. In this case, X is strictly stationary.

In addition to the definitions in Expression (3.2), we require the fourth-order moments of X, which are given by

$$E(X_t X_{t+\tau} X_s X_{s+\nu}) = \sum_{i,j,k,l=0}^{\infty} a_i a_j a_k a_l E(\epsilon_{t-i}\epsilon_{t+\tau-j}\epsilon_{s-k}\epsilon_{s+\nu-l}).$$

The four nonzero product summations occur for

$$t - i = t + \tau - j = s - k = s + \nu - l,$$
$$t - i = t + \tau - j \neq s - k = s + \nu - l,$$
$$t - i = s - k \neq t + \tau - j = s + \nu - l,$$
$$t - i = s + \nu - l \neq t + \tau - j = s - k,$$

so that

$$\begin{aligned}
E(X_t X_{t+\tau} X_s X_{s+\nu}) &= \varkappa_{s-t,\tau,\nu} + \sum_{i=0}^{\infty} a_i a_{i+\tau} \sum_{k=0}^{\infty} a_k a_{k+\nu} \\
&+ \sum_{i=0}^{\infty} a_i a_{i-t+s} \sum_{j=0}^{\infty} a_j a_{j-t+s-\tau+\nu} \\
&+ \sum_{i=0}^{\infty} a_i a_{i-t+s+\nu} \sum_{j=0}^{\infty} a_j a_{j-t+s-\tau} \\
&= \varkappa_{s-t,\tau,\nu} + R_\tau R_\nu + R_{t-s} R_{t-s+\tau-\nu} \\
&+ R_{t-s-\nu} R_{t-s+\tau},
\end{aligned}$$

the fourth-order cumulant being

$$\varkappa_{s-t,\tau,\nu} = \varkappa \sum_{i=0}^{\infty} a_i a_{i+\tau} a_{i-t+s} a_{i-t+s+\nu}.$$

Notice that this cumulant is a function of the difference $(s - t)$, not of s and t separately.

3.2. Autocorrelation Analysis

Suppose we observe the sequence X without error at T unit intervals. The most commonly employed sample autocovariance function is

$$(3.3) \qquad C_\tau = N^{-1} \sum_{t=1}^{T-\tau} X_t X_{t+\tau},$$

where the divisor N is defined as $(T - \tau)$ or T. When the quantity $(T - \tau)$ is used, C is an unbiased estimator of the autocovariance function R. When N is the sample record length T, then a negative bias,

$$E(C_\tau - R_\tau) = -\tau R_\tau / T,$$

occurs. Parzen has pointed out two desirable properties of the biased estimator [73, p. 174]. Firstly it leads to a smaller mean-square error for a given sample record length T than does the unbiased estimator. While no theorem to this effect currently exists, Shaerf [86] has shown that a number of commonly employed covariance stationary models all lead to this result. Secondly the biased estimator is positive definite whereas the unbiased estimator is not. As Section 3.4 will show, the biased estimator (when used in a particular way) leads to positive estimates of the spectrum; the unbiased estimator does not necessarily do so.

While these properties are desirable for theoretical reasons, statisticians generally agree that the choice of divisor makes little difference for spectrum estimation in most circumstances when the time series are sufficiently long [48, 90]. This is especially true when T is considerably greater than τ. In keeping with the growing precedent, we use the biased estimator in the remainder of this study.† Besides the properties just mentioned, using the biased estimator simplifies the derivation of the sampling properties.

Regardless of the choice of divisor, the estimator C_τ is a consistent estimator of R_τ for fixed τ. The sample autocovariance function C, however, is not a consistent estimator of the autocovariance function R, since the integrated mean square error tends to a positive limit as T becomes large [72]. This shortcoming does not bear directly on the present description. It should be noted in passing that Shaerf has considered a more general class of estimators than that of Expression (3.3), but her results do not appear encouraging [86].

---

† This practice has been followed by Grenander and Rosenblatt [32] and in the works of Jenkins and Parzen.

Most time series in economics have unknown nonzero means; therefore Expression (3.3) is inappropriate. Two new problems must now be resolved. One is to choose an estimator of the mean, and the other is to incorporate it into the sample autocovariance function. The most commonly used sample mean is

$$\bar{X} = T^{-1} \sum_{t=1}^{T} X_t.$$

We replace each X_t in Expression (3.3) with its corresponding $(X_t - \bar{X})$ so that

(3.4) $$C_\tau = T^{-1} \sum_{t=1}^{T-\tau} [(X_t - \bar{X})(X_{t+\tau} - \bar{X})].$$

Statisticians, primarily those concerned with correlation analysis, often prefer to use an estimator having the form

(3.5) $$C_\tau = T^{-1} \left[\sum_{t=1}^{T-\tau} X_t X_{t+\tau} - (T - \tau)^{-1} \left(\sum_{t=1}^{T-\tau} X_t \right) \left(\sum_{t=1}^{T-\tau} X_{t+\tau} \right) \right].$$

The latter choice is considered more appropriate when the time series is short and the end points are appreciably greater or less than the mean.† In spectrum analysis, the time series are of sufficient length to remove the sensitivity to end-point values, especially after pre-whitening has been performed.‡ In addition we consider large sample properties, so that the choice of estimators then makes little difference. We therefore use Expression (3.4), which is commonly used in spectrum analysis.*

Using Expression (3.4), we have§

$$E(C_\tau) = T^{-1} \left[(T - \tau)R_\tau + \frac{(T - \tau)}{T^2} \sum_{t,s=1}^{T} R_{t-s} \right.$$

$$\left. - T^{-1} \sum_{t=1}^{T-\tau} \sum_{s=1}^{T} (R_{t-s} + R_{t+\tau-s}) \right]$$

$$= T^{-1} \left[(T - \tau)R_\tau + \frac{(T - \tau)}{T} \sum_{v=-(T-1)}^{T-1} (1 - |v|/T) R_v \right.$$

$$\left. - \sum_{v=-(T-1)}^{T-\tau-1} \phi_v(R_v + R_{v+\tau}) \right], \qquad \tau \geq 0,$$

---

† See Marriott and Pope [64, p. 391], and Weinstein [98, p. 882].
‡ See Section 3.11.
* See Blackman and Tukey [11, p. 53] and Parzen [75, pp. 939–940].
§ The corresponding bias for Expression (3.5) may be found in Kendall [53, p. 403].

where

$$\phi_v = \begin{cases} 1 + v/T, & -(T-1) \le v \le -\tau \\ 1 - \tau/T, & -\tau \le v \le 0 \\ 1 - v/T, & 0 < v \le T - \tau - 1. \end{cases}$$

Since R is a function of bounded variation, one may show that

$$\lim_{T \to \infty} TE(C_\tau - R_\tau) = -\left(\tau R_\tau + \sum_{v=-\infty}^{\infty} R_v\right) = -[\tau R_\tau + 2\pi g(0)].$$

When $T \gg \tau$, the first term of the asymptotic bias, $-\tau R_\tau/T$, is negligible compared with $-2\pi g(0)/T$ so that, for large T,

$$E(C_\tau - R_\tau) \sim -2\pi g(0)/T,$$

which is always negative. The bias has the same limiting form irrespective of the choice of sample mean in the sample autocovariance function.

Unlike the case of independent observations, the bias cannot be eliminated unless $g(0)$ is known. In addition, the bias is uniform for all τ. Since R will generally be a decreasing function, the relative bias increases with τ. Later we shall see that bias due to the sample mean emerges in the sample spectrum principally in the low-frequency range.

In the second example of Section 2.11 we observed that for a damped periodic autocovariance function,

$$f(0) = \alpha[\pi(\alpha^2 + \lambda_0^2)]^{-1}.$$

This implies that the higher the frequency λ_0, the smaller the bias for a given damping constant α. This result is intuitively plausible because the higher the frequency of oscillation, the more often the mean is crossed per unit time. Positive deviations are then more readily compensated for by negative deviations. From the third example of that section we observe that additional periodic components also reduce the bias.

For convenience of exposition we again assume a zero mean and use the estimator in Expression (3.3) to derive other sampling properties. Later we shall introduce the sample mean bias where appropriate. The covariance of the estimator C_τ is then

$$\text{cov}(C_\tau, C_v) = T^{-2} \sum_{t=1}^{T-\tau} \sum_{s=1}^{T-v} \text{cov}(X_t X_{t+\tau}, X_s X_{s+v})$$

$$= T^{-2} \sum_{t=1}^{T-\tau} \sum_{s=1}^{T-v} [(R_{t-s}R_{t-s+\tau-v} + R_{t-s-v}R_{t-s+\tau})$$

$$+ \varkappa_{s-t,\tau,v}]$$

$$= T^{-2} \sum_{v=-(T-v-1)}^{T-\tau-1} \{w_v[(R_v R_{v+\tau-v} + R_{v-v}R_{v+\tau})$$

$$+ \varkappa_{-v,\tau,v}]\},$$

where

$$w_v = \begin{cases} T - v - v, & 1 - T + v \le v \le 0 \\ T - v, & 0 \le v \le v - \tau \\ T - \tau - v, & v - \tau \le v \le T - \tau - 1, \end{cases} \quad \tau \ge v.$$

In particular, the variance of the estimator C is then

$$\text{var}(C_\tau) = T^{-2} \sum_{v=-(T-1)}^{T-1} \{(T - |v|)[(R_v^2 + R_{v+\tau}R_{v-\tau}) + \varkappa_{-v,\tau,\tau}]\}.$$

As T becomes large, we have†

$$\lim_{T \to \infty} T \, \text{cov}(C_\tau, C_v) = \sum_{v=-\infty}^{\infty} (R_v R_{v+\tau-v} + R_{v+\tau}R_{v-v}) + \varkappa R_\tau R_v$$

$$= 4\pi \int_{-\pi}^{\pi} g^2(\omega) \cos \omega\tau \cos \omega v \, d\omega + \varkappa R_\tau R_v.$$

There are two sampling properties that are of interest here. One is the asymptotic covariance between sample covariances, which is given by

$$(3.6) \quad \text{cov}(C_\tau, C_v) \sim T^{-1} \left[\sum_{v=-\infty}^{\infty} (R_v R_{v+\tau-v} + R_{v+\tau}R_{v-v}) + \varkappa R_\tau R_v \right].$$

The other is the asymptotic variance given by

$$(3.7) \quad \text{var}(C_\tau) \sim T^{-1} \left[\sum_{v=-\infty}^{\infty} (R_v^2 + R_{v+\tau}R_{v-\tau}) + \varkappa R_\tau^2 \right].$$

We observe that, even for large samples, the estimated covariances are correlated. Any unusual sampling fluctuation in one estimate influences the remaining ones, and the extent of this influence depends on the degree of correlation in the original sequence and on $\varkappa$ for a non-normal sequence. This, however, is the very information we seek to determine. We also observe that, for large τ, the large sample variance is essentially

$$T^{-1} \sum_{v=-\infty}^{\infty} R_v^2.$$

This implies that, to a reasonable approximation, the magnitude of sampling fluctuations for a given τ is unrelated to the magnitude of R_τ.

When the sample mean is used in place of the true mean, stability is more appropriately measured by the mean-square error than by the

---

† See Bartlett [5, pp. 253–256].

variance alone. In the current case, the squared bias is of order T^{-2}, whereas the variance is of order T^{-1}. For large T, there is little difference between the two stability measures.

In correlation analysis, statistical inference is based on the *correlogram*, which is a graph of the sample autocorrelation function† $r_\tau = C_\tau/C_0$. Using a Taylor series expansion in two variables, we have to order T^{-1}

$$(3.8) \quad E(r_\tau) = E(C_\tau)/E(C_0) - \mathrm{cov}\,(C_\tau, C_0)/[E(C_0)]^2 \\ + E(C_\tau)\,\mathrm{var}\,(C_\tau)/[E(C_0)]^3.$$

We observe that r_τ is a biased estimator of the correlation coefficient ρ_τ. The bias arises from four sources: the use of a ratio estimator, the choice of divisor, the substitution of the sample mean, and the inherent correlation between the numerator and denominator in the ratio estimator.

Sampling investigations by Marriott and Pope [63] have shown bias to be a serious problem for relatively simple autocorrelation structures and large T. Kendall [53] reached the same conclusion and has noted that when autocorrelation is appreciable, terms to the order T^{-2} become relevant in Expression (3.8).

If the variation of C_τ around R_τ is reasonably small, we have to order T^{-1}

$$\mathrm{cov}\,(r_\tau, r_\nu) = \frac{\mathrm{cov}\,(C_\tau, C_\nu)}{[E(C_0)]^2} - \frac{E(C_\tau)\,\mathrm{cov}\,(C_\nu, C_0)}{[E(C_0)]^3} \\ - \frac{E(C_\nu)\,\mathrm{cov}\,(C_\tau, C_0)}{[E(C_0)]^3} + \frac{E(C_\tau)E(C_\nu)\,\mathrm{var}\,(C_0)}{[E(C_0)]^4}.$$

Using Expression (3.6) we have for large T

$$(3.9a) \quad \mathrm{cov}\,(r_\tau, r_\nu) \sim T^{-1} \sum_{v=-\infty}^{\infty} [\rho_v\rho_{v+\tau-\nu} + \rho_{v+\tau}\rho_{v-\nu} - 2(\rho_\tau\rho_v\rho_{v+\nu} \\ + \rho_\nu\rho_v\rho_{v+\tau} - \rho_\tau\rho_\nu\rho_v^2)],$$

$$(3.9b) \quad \mathrm{var}\,(r_\tau) \sim T^{-1} \sum_{v=-\infty}^{\infty} [\rho_v^2 + \rho_{v+\tau}\rho_{v-\tau} - 2(2\rho_\tau\rho_v\rho_{v+\tau} - \rho_\tau^2\rho_v^2)].$$

Notice that terms proportional to $\varkappa$ vanish algebraically.

While it is possible to derive analytical expressions for bias, covariation, and variance for correlogram estimates, their dependence

---

† Other forms of the estimator are discussed in Marriott and Pope [63] and Weinstein [98]. The latter paper considers alternative estimators for small samples.

on the true correlation structures makes them somewhat inadequate for gauging the value of estimates. In addition, the correlation between estimates can easily obscure the true appearance of the autocorrelation function and consequently one must interpret the correlogram with caution.

With regard to the relevant distribution theory, Walker [95], Lomnicki and Zaremba [61], and Parzen [74] have stated alternative conditions that lead to the joint distribution of the sample autocovariances being asymptotically normal. If X is a discrete linear process, then joint asymptotic normality holds for the sample autocovariances and, in addition, it holds for the sample autocorrelation coefficients. To use these results, the covariance between estimates must be known; this again depends on the very function being estimated.†

3.3. Spectrum Averaging

Before turning to spectrum estimation, we shall find it instructive to study some peculiarities of frequency-domain analysis that distinguish it from more traditional analyses. These peculiarities result from using time series of finite length together with the transformation of a time function into a frequency function. If $g(\lambda)$ represents the spectrum at angular frequency λ, and the stochastic sequence X is observed at unit intervals, then we have

$$(3.10a) \qquad g(\lambda) = (2\pi)^{-1} \sum_{\tau=-\infty}^{\infty} R_\tau e^{-i\lambda\tau},$$

$$(3.10b) \qquad R_\tau = \int_{-\pi}^{\pi} g(\lambda)e^{i\lambda\tau} \, d\lambda \qquad \tau = 0, \pm 1, \pm 2, \ldots, \pm \infty.$$

Now consider applying a *weighting function* k to the autocovariance function, such that

$$(3.11) \qquad g'(\lambda) = (2\pi)^{-1} \sum_{\tau=-\infty}^{\infty} k_{M,\tau} R_\tau e^{-i\lambda\tau},$$

where M is an arbitrary constant and $k_{M,\tau} = k_{M,-\tau}$. Using Expression (3.10b), the function g' may be rewritten as

$$g'(\lambda) = (2\pi)^{-1} \sum_{\tau=-\infty}^{\infty} \left[k_{M,\tau} e^{-i\lambda\tau} \int_{-\pi}^{\pi} g(\omega)e^{i\omega\tau} \, d\omega \right]$$

$$= (2\pi)^{-1} \int_{-\pi}^{\pi} g(\omega) \left[\sum_{\tau=-\infty}^{\infty} k_{M,\tau} e^{i(\omega-\lambda)\tau} \right] d\omega.$$

† For a more detailed discussion, see Parzen [74].

Let

(3.12) $$K_M(\lambda) = (2\pi)^{-1} \sum_{\tau=-\infty}^{\infty} k_{M,\tau} e^{-i\lambda\tau}$$

so that

$$g'(\lambda) = \int_{-\pi}^{\pi} g(\omega) K_M(\lambda - \omega) \, d\omega.$$

The spectral function g' is a weighted average of the true spectrum where the *averaging kernel* or *spectral window* K is the Fourier transform of the *weighting function* or *lag window* k. To understand the significance of this averaging procedure, we consider a simple, yet very important example. Since time series are of finite length T, we can estimate at most $T - 1$ autocovariances. Suppose that we actually know the first M autocovariances with certainty. Our natural instinct would be to replace the upper limit of summation in Expression (3.11) by M. This corresponds to choosing a weighting function

$$k_{M,\tau} = \begin{cases} 1, & \tau = 0, \pm 1, \pm 2, \ldots, \pm M, \\ 0, & \text{elsewhere,} \end{cases}$$

so that

$$g'(\lambda) = (2\pi)^{-1} \sum_{\tau=-M}^{M} R_\tau e^{-i\lambda\tau}.$$

The corresponding averaging kernel is then

(3.13) $$K_M(\lambda - \omega) = \frac{\sin\left[(M + \tfrac{1}{2})(\lambda - \omega)\right]}{2\pi \sin\left[(\lambda - \omega)/2\right]},$$

which permits us to compute the spectrum average g'—not the spectrum g.

Figure 17 shows the averaging kernel corresponding to Expression (3.13). Observe the sizeable side lobes, the first of which is negative and more than 20 percent of the main lobe. Now, as M increases, the averaging kernel becomes more concentrated at frequency λ, and the spectrum average $g'(\lambda)$ converges to the spectrum ordinate $g(\lambda)$. Unfortunately, this averaging kernel converges slowly so that, even with large M, a considerable amount of distortion is present in $g'(\lambda)$. This distortion, commonly called *leakage*, comes from other frequency bands and leads to a nonrepresentative spectrum average of the spectrum.

Because of the poor convergence properties of this weighting function, it would appear desirable to use all $T - 1$ sample auto-covariances in the spectrum estimator to get the best convergence

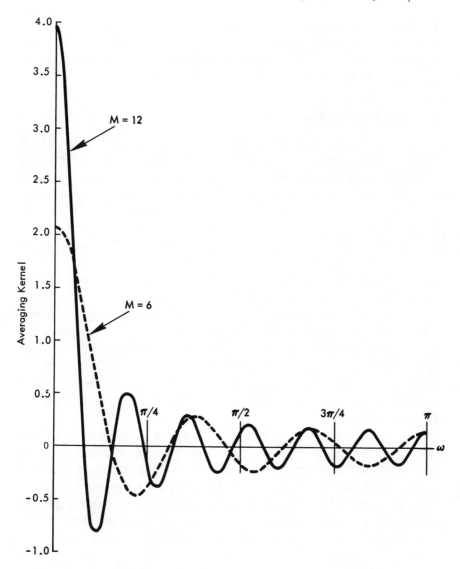

Fig. 17 Averaging kernel corresponding to the unit weighting function

possible. Unfortunately, this procedure causes the resulting estimator to lack the statistical property of consistency. This inadequacy will be described in detail in Section 3.6.

It is generally true that as long as the weighting function k, regardless of its form, uses a finite number of autocovariances, the corresponding spectral representation at a given frequency is a spectrum average over the spectrum. Now it is possible to speed up the averaging kernel's rate of convergence by an appropriate choice of weighting function k and, in addition, to insure the consistency of the spectrum estimator. Our analysis must therefore deal with two problems, one of approximation, the other of estimation.

3.4. Averaging Kernels

It should now be clear to the reader that a unit weighting of a finite number of autocovariances does not guarantee that the resulting spectrum averages will resemble the spectrum. This fact is independent of statistical questions that must also be considered when evaluating the adequacy of spectrum estimates. For the present we concentrate on spectrum averaging, assuming that the autocovariance function is known. Once the necessity for appropriate spectrum averaging is brought into perspective, we then consider the statistical properties of spectrum estimates. In this way we first concentrate on an appropriate *approximation*, and then study the estimation problem.

We begin by describing the properties that a weighting function k and a corresponding averaging kernel K should have for representative spectrum averaging. In spectrum analysis our purpose is to estimate the spectrum within a frequency band around a given frequency. Apart from variance considerations we wish to make this band as small as possible in order to study the shape of spectrum as a function of frequency.

We wish to use a weighting function whose corresponding averaging kernel has small side lobes compared to its main lobe and which concentrates its main lobe around the point of interest. Small side lobes reduce the possibility of distortion from remote frequencies. Concentrating the main lobe around the principal frequency gives the greatest weight to its corresponding spectrum ordinate. To accomplish the first requirement, the weighting function k should be smooth and slowly changing.† To accomplish the second requirement, the weighting function should be just the opposite—flat and blocky. These requirements follow from the nature of weighting functions and their Fourier transforms. Since these two desirable properties work in opposite directions, some compromise is necessary.

Many weighting functions have been discussed in the literature, and their properties have been summarized in review articles by Jenkins [47] and Parzen [73]. Table 4 lists the two most commonly employed

---

† See Blackman and Tukey [11, p. 14].

Table 4

Most Commonly Employed Weighting Functions and Their Corresponding Averaging Kernels

Originator	$k_{M,\tau}$		$k_M(\omega)$
Tukey-Hanning	$\frac{1}{2}[1 + \cos(\pi\tau/M)]$	$\tau < M$	$\frac{1}{2\pi}\left\{\dfrac{\sin(M+\frac{1}{2})\omega}{\sin \omega/2} + \dfrac{1}{2}\left[\dfrac{\sin(M+\frac{1}{2})(\omega+\pi/M)}{\sin\frac{1}{2}(\omega-\pi/M)}\right.\right.$
	0	$\tau \geq M$	$\left.\left.+\ \dfrac{\sin(M+\frac{1}{2})(\omega-\pi/M)}{\sin\frac{1}{2}(\omega-\pi/M)}\right]\right\}$
Parzen	$1 - 6|\tau/M|^2 + 6|\tau/M|^3$	$\tau \leq M/2$	$\dfrac{3}{8\pi M^3}\left[\dfrac{\sin(M\omega/4)}{\sin(\omega/4)}\right]^4$
	$2(1 - |\tau/M|)^3$	$M/2 < \tau < M$	
	0	$\tau \geq M$	

Source: Jenkins [47].

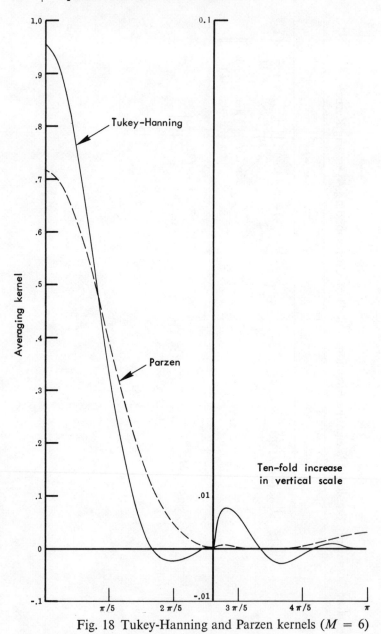

Fig. 18 Tukey-Hanning and Parzen kernels ($M = 6$)

weighting functions, and Fig. 18 shows the corresponding averaging kernels for M equal to 6 lags. Note that the side lobes of both spectral windows are small compared to the main lobes. Observe also that Parzen's window is nonnegative everywhere, while the Tukey-Hanning one is not. If Parzen's weighting function is used with the biased sample autocovariance function in Expression (3.4), the resulting estimates will necessarily be nonnegative.† This is not true of the Tukey-Hanning weighting function, and so it is possible to have negative spectrum estimates with this weighting function regardless of whether the biased or unbiased sample autocovariance function is used.

The fact that one window leads to nonnegative estimates and the other does not is insufficient reason for choosing the former over the latter. The reader should bear in mind that we are estimating a function, not a point. Therefore our concern is with the overall shape of the estimated function—not with the point estimates themselves. If negative estimates do occur, they need only be judged in comparison with the remaining point estimates of the spectrum.

Note also that negative estimates obtained with the Tukey-Hanning window indicate that leakage is a serious problem in at least some frequency bands. Using this window therefore provides a warning of serious bias due to leakage.

We have already stressed the fact that an averaging kernel should concentrate its main lobe around the frequency of interest in order for us to obtain spectrum averages that are representative of the spectrum. This degree of concentration or focusing power is customarily measured by *bandwidth*, which may be defined in several ways [47, 73]. For current purposes, we define the bandwidth β as the width of a rectangle whose height and area correspond to those of the averaging kernel at the frequency of interest. So bandwidth is

$$\beta_M = K_M^{-1}(0) \int_{-\pi}^{\pi} K_M(\omega) \, d\omega,$$

and is measured in radians. The area under the averaging kernel is usually unity so that

$$\beta_M = K_M^{-1}(0) = 2\pi \left(\sum_{\tau=-M}^{M} k_{\tau, M} \right)^{-1}.$$

Column (1) of Table 5 lists the bandwidth of the three aforementioned averaging kernels. Observe that for a given number of lags the Tukey-Hanning window has a bandwidth 77 percent that of Parzen's. One may prematurely conclude from this observation that the former window is better than the latter. We cannot make any judgment, however, until we examine the variances of the respective

† This follows from the positive definiteness of the biased autocovariance estimator.

Table 5

Bandwidth, Variance, and Equivalent Degrees of Freedom
for Three Averaging Kernels[a]

Averaging Kernel	(1) Bandwidth	(2) Variance/$g^2(\lambda)$	(3) EDF
Unit	$\pi/(M + \frac{1}{2})$	2.00 M/T	T/M
Tukey-Hanning	$2\pi/M$	0.75 M/T	$2.7T/M$
Parzen	$8\pi/(3M)$	0.54 M/T	$3.7T/M$

[a]At frequencies zero and $\pm\pi$, the variance is doubled and the EDF is halved. The variance and EDF are for a normal process.

estimators. Note also that the unit weighting function leads to a smaller bandwidth than either of the other windows. This is due to its blocky nature, which concentrates the main lobe more closely around the central frequency. Unfortunately the blockiness also induces large side lobes. Here it becomes clear that bandwidth by itself cannot be the only criterion for measuring focusing power since, as Fig. 17 shows, the unit weighting function does not result in representative averaging. The shape of an averaging kernel plays an important role in determining its desirability.

If the spectrum changes slowly over the bandwidth, then we consider our spectrum averages to be representative of the spectrum. The averages are said to be *well resolved*. It is important to note that specifying the bandwidth does not insure against leakage and thereby guarantee good resolution. The shape of the spectrum in the interval $(\lambda - \beta_M/2, \lambda + \beta_M/2)$ determines how good the resolution of a given spectrum average is. The slope of the spectrum may vary significantly over the frequency domain and, therefore, the number of lags needed to resolve the spectrum at one point may be insufficient to resolve it at another point. This fact may sometimes necessitate using different numbers of lags when estimating different parts of the spectrum.

To check for good resolution one may estimate the spectrum for several values of M. In the absence of sampling fluctuations the spectrum average centered on λ would converge mathematically to $g(\lambda)$ as M increases. This follows from averaging over smaller and smaller intervals in which the spectrum changes less and less. Since sampling fluctuations are present, the convergence will not be as apparent as it would be if we were averaging over the true spectrum. Notice that using several values of M reduces the advantage of the Tukey-Hanning window over the Parzen window as an indicator of serious leakage.

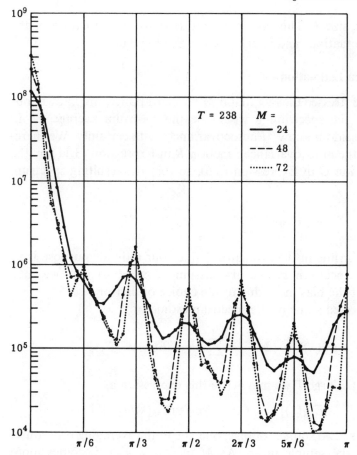

Fig. 19 Sample spectrum of manufacturers' shipments of durable goods, seasonally unadjusted

Figure 19 shows sample spectra of seasonally unadjusted shipments for 24, 48, and 72 lags. All spectra were estimated using the Parzen window. For M equal to 24, the curve appears smooth. The seasonal peaks are much sharper for 48 lags than for 24. The curve using 72 lags has even sharper peaks, but the resolution does not seem greatly improved over the 48 lag curve. We conclude that 48 lags suffice to give good resolution.

This method of determining good resolution is clearly subjective. The ideal approach would be to have a test statistic that would determine the appropriate M. Zaremba [109] discusses this issue in some detail and points out the difficulties involved in deriving such a statistic.

There remains the question of how large an M is acceptable for a given sample size T. This consideration depends on the nature of the sampling fluctuations and is discussed in Section 3.6.

3.5. Spectrum Estimation

We have stressed the fact that if M autocovariances are known, our estimator of the spectrum g would be the spectrum average g'. In practice, we seldom know M autocovariances with certainty. We therefore replace the autocovariance function R in Expression (3.11) by the sample function C in Expression (3.6), so that the resulting spectrum estimator is

$$(3.14) \qquad \hat{g}(\lambda) = (2\pi)^{-1} \sum_{\tau=-M}^{M} k_{M,\tau} C_\tau e^{-i\lambda\tau}.$$

We omit the prime in the estimator $\hat{g}$ since our ultimate purpose is to estimate the spectrum g. We also assume $T \gg M$ so that we may safely neglect the bias in C_τ due to the choice of divisor T.

The expected value of the spectrum estimator is

$$E[\hat{g}(\lambda)] = (2\pi)^{-1} \sum_{\tau=-M}^{M} k_{M,\tau} E(C_\tau) e^{-i\lambda\tau}.$$

For sufficiently large T we may write this expression as

$$E[\hat{g}(\lambda)] \sim g'(\lambda) - 2\pi g(0) K_M(\lambda)/T,$$

which contains two sources of bias, one due to the averaging procedure, the other to the sample mean. As M increases, $g'(\lambda)$ becomes more concentrated around frequency λ so that $g'(\lambda) \sim g(\lambda)$. The rate of convergence clearly depends on the choice of averaging kernel. For current purposes it is sufficient to note this asymptotic convergence with increasing M. In Section 3.8, we discuss the effects of the averaging bias on the mean-square error. In the remainder of the study we assume this bias to be negligible.†

The remaining negative bias is due to the substitution of the sample mean and is affected by both λ and M. For a given M, the averaging kernel K has a maximum at frequency zero and decreases with increasing $|\lambda|$. Since bandwidth measures the interval over which K concentrates virtually all of its area, we expect K to be negligible beyond frequencies $\pm\beta_M/2$.

In the neighborhood of frequency zero, K increases with increasing M. Away from zero, K decreases. It becomes apparent that bias due to

---

† Bias due to averaging is discussed more fully in Hannan [36, pp. 63–64].

the sample mean primarily affects the extreme low-frequency range in contrast to its uniform influence on the sample autocovariance function. This property is especially convenient because we can compensate for the bias by noting that

$$E[\hat{g}(0)] \sim g(0)[1 - 2\pi K_M(0)/T].$$

An unbiased estimator of $g(0)$ is then

$$\hat{g}_u(0) = \hat{g}(0)[1 - 2\pi K_M(0)/T]^{-1},$$

so that

$$E[\hat{g}_u(0)] \sim g(0).$$

We may then write

$$\hat{g}_u(\lambda) = \hat{g}(\lambda) + 2\pi K_M(\lambda)\hat{g}_u(0)/T,$$

for which

$$E[\hat{g}_u(\lambda)] \sim g(\lambda).$$

As mentioned earlier, this correction need only be made for estimates in the interval near zero frequency. We see that sample mean bias can be easily dealt with in the frequency domain. In the remainder of the chapter we assume this correction to be made and drop the subscript u.

We ponder the sample mean bias in the frequency domain with good cause. Failure to recognize its presence may lead us to interpret a low-frequency peak as a genuine peak corresponding to some reasonably regular cyclical phenomenon in the time domain. Since the bias is negative, it reduces the contribution nonuniformly at very low frequencies and may thus create a spurious peak in the spectrum. This point should be remembered when interpreting the estimated spectrum.

3.6. Covariance Properties of Spectrum Estimates

The covariance function for spectrum estimates is

(3.15)

$$\text{cov}\,[\hat{g}(\lambda), \hat{g}(\lambda')] = (2\pi)^{-2} \sum_{\tau,\nu=-M}^{M} k_{M,\tau} k_{M,\nu}\, \text{cov}\,(C_\tau, C_\nu) e^{-i(\lambda\tau + \lambda'\nu)}.$$

Using the limiting expression for $\text{cov}\,(C_{T,\tau}, C_{T,\nu})$, we have

$$\lim_{T \to \infty} T\,\text{cov}\,[\hat{g}(\lambda), \hat{g}(\lambda')] = \pi \int_{-\pi}^{\pi} g^2(\omega)[K_M(\omega - \lambda) + K_M(\omega + \lambda)]$$

$$\times [K_M(\omega - \lambda') + K_M(\omega + \lambda')]\,d\omega + \varkappa g'(\lambda)g'(\lambda').$$

Asymptotically we have

$$\lim_{M \to \infty} \lim_{T \to \infty} M^{-1}T \operatorname{cov}[\hat{g}(\lambda), \hat{g}(\lambda')]$$

$$= \begin{cases} cg^2(\lambda), & \lambda = \pm\lambda' \neq 0, \pm\pi \\ 2cg^2(\lambda), & \lambda = \lambda' = 0, \lambda = \pm\lambda' = \pm\pi \\ 0, & \lambda \neq \lambda', \end{cases}$$

where

$$c = \begin{cases} .75 & \text{Tukey-Hanning} \\ .54 & \text{Parzen.} \end{cases}$$

For sufficiently large T and M, the variance of a spectrum estimate is approximately

$$(3.16) \qquad \operatorname{var}[\hat{g}(\lambda)] \sim (\delta_\lambda \Psi_{T,M} + \varkappa/T)g^2(\lambda),$$

$$(3.17) \qquad \Psi_{T,M} = 2\pi T^{-1} \int_{-\pi}^{\pi} K_M^2(\omega)\, d\omega = T^{-1} \sum_{\tau=-M}^{M} k_{M,\tau}^2 = cM/T,$$

$$\delta_\lambda = \begin{cases} 2, & \lambda = 0, \pm\pi \\ 1, & \lambda \neq 0, \pm\pi. \end{cases}$$

The variances for a normal process are listed in column (2) of Table 5. Suppose we compute estimates using all three kernels with equal bandwidth. The variance of the Parzen estimates is then about 0.96 that of the Tukey-Hanning estimates. This implies that both kernels give similar results for bandwidth and variance, albeit for different M's. With regard to the estimates using the unit averaging kernel, we note that they have a variance about four-thirds that of the Tukey-Hanning and Parzen estimates for the same bandwidth. Throughout this study we shall use the Parzen window.

While more lags improve resolution, they also increase the variance of a given sample record length T. As T increases it is, of course, reasonable to increase M. The relative increase in M, however, should be smaller than that in T, so that we may improve the stability of our estimates at the same time that we are improving resolution. The point to bear in mind is that stability is inversely proportional to the ratio M/T.

Table 6 may be used to gauge the effects of M/T on stability. If, for example, resolution is reasonably accomplished using the Tukey-Hanning window with M/T equal to 0.20, we observe that the standard deviation of our estimates is roughly 40 percent of the mean. This is clearly an underestimate if our sequence is nonnormal, for then $\varkappa$

Table 6

Approximate Variances and Standard Deviations as Proportions of $g^2(\lambda)$ and $g(\lambda)$, Respectively, for a Normal Process

M/T	Tukey-Hanning Window		Parzen Window	
	$\Psi_{T,M}$	$(\Psi_{T,M})^{\frac{1}{2}}$	$\Psi_{T,M}$	$(\Psi_{T,M})^{\frac{1}{2}}$
0.3	0.2250	0.47	0.1626	0.40
0.25	0.1875	0.43	0.1355	0.37
0.20	0.1500	0.39	0.1084	0.33
0.15	0.1125	0.34	0.0813	0.28
0.10	0.0750	0.27	0.0542	0.23
0.05	0.0375	0.19	0.0271	0.16
0.02	0.0150	0.12	0.0108	0.10
0.01	0.0075	0.09	0.0054	0.07

is nonzero and must be taken into account. Notice also that the choice of M/T depends on the windows we are using, since these affect resolution and stability. It is clear that regardless of the window choice, a ratio greater than 0.3 will not give very reliable results in either case.

3.7. Correlation between Spectrum Estimates

From Expression (3.16) we note that spectrum estimates are asymptotically uncorrelated for normal processes. It is instructive now to study the correlation between estimates for finite M and T, since this is the case commonly encountered. Suppose X were a white noise normal process. Then the correlation between estimates is easily seen to be

$$(3.18) \quad \text{corr}\,[\hat{g}(\lambda),\hat{g}(\lambda')]$$

$$= \left(\sum_{\tau=-M}^{M} k_{M,\tau}^2 \cos \lambda\tau \cos \lambda'\tau \right) \left(\sum_{\tau=-M}^{M} k_{M,\tau}^2 \cos^2 \lambda\tau \right)^{-1}.$$

Table 7 shows the correlation for both windows for separation increments of $\pi/(10M)$ radians. Notice that Parzen's window has higher correlation for each separation. This is to be expected, since its bandwidth is greater for a given M. The arrows in the table indicate the correlations at bandwidth separations, and we note that the correlations at these points are small and not much different for the two kernels. For

Table 7

Correlation between Spectrum Estimates for a White Noise
Normal Process[a]

Separation $(n\pi/M)$	Correlation		Separation $(n\pi/M)$	Correlation	
n	Tukey-Hanning	Parzen	n	Tukey-Hanning	Parzen
0.0	1.000	1.000	2.6	0.0292	0.2100
0.1	0.9961	0.9978	2.7	0.0184	→0.1850
0.2	0.9843	0.9911	2.8	0.0101	0.1620
0.3	0.9650	0.9802	2.9	0.0041	0.1410
0.4	0.9385	0.9650	3.0	0.0000	0.1219
0.5	0.9054	0.9458	3.1	−0.0026	0.1048
0.6	0.8663	0.9228	3.2	−0.0040	0.0894
0.7	0.8220	0.8964	3.3	−0.0045	0.0758
0.8	0.7734	0.8667	3.4	−0.0044	0.0637
0.9	0.7213	0.8342	3.5	−0.0038	0.0533
1.0	0.6667	0.7993	3.6	−0.0030	0.0442
1.1	0.6105	0.7623	3.7	−0.0021	0.0365
1.2	0.5537	0.7236	3.8	−0.0013	0.0300
1.3	0.4972	0.6837	3.9	−0.0006	0.0246
1.4	0.4417	0.6429	4.0	0.0000	0.0202
1.5	0.3881	0.6017	4.1	0.0004	0.0166
1.6	0.3369	0.5605	4.2	0.0006	0.0137
1.7	0.2888	0.5196	4.3	0.0006	0.0114
1.8	0.2442	0.4794	4.4	0.0006	0.0096
1.9	0.2034	0.4401	4.5	0.0005	0.0083
2.0	→0.1667	0.4020	4.6	0.0003	0.0073
2.1	0.1340	0.3655	4.7	0.0002	0.0065
2.2	0.1054	0.3306	4.8	0.0001	0.0060
2.3	0.0809	0.2975	4.9	0.0000	0.0057
2.4	0.0602	0.2663	5.0	0.0000	0.0057
2.5	0.0431	0.2371			

[a]Correlations were computed using Expression (3.18), with $\lambda = \pi/2$, $\lambda' = \lambda - n\pi/M$, and $M = 10$.

purposes of analysis, it is reasonable to consider estimates more than a bandwidth apart as uncorrelated.

Suppose that the spectrum changes slowly over the bandwidth. Then the correlations within a bandwidth hold approximately, since virtually all the overlap between averaging kernels occurs within this interval. We may then regard estimates more than a bandwidth apart as uncorrelated for practical purposes.

In theory, we may estimate as many points as we wish, bearing in mind that correlation is inversely related to the separation between estimates. To derive an adequate representation of the spectrum, we will often compute estimates less than a bandwidth apart. The point of significance is that we can determine the separation between reasonably uncorrelated estimates, a feature absent in correlation analysis where a knowledge of the autocovariance function is necessary. Estimating the spectrum at frequency increments of π/M generally provides a representative picture of the sample spectrum and also avoids the need for an additional design parameter.

3.8. Criteria for Choosing among Estimators

Since we are estimating a function and not a point, it seems appropriate to judge function estimators by a criterion that takes into account the deviations over the domain of the function. One logical criterion to apply is the integrated mean-square error of the estimator, mentioned in Section 3.2; the integrated mean-square errors for both biased and unbiased estimators of the autocovariance function converge to positive limits as the sample record length becomes large [72]. This fact induces some skepticism about the use of the sample autocovariance function; it is therefore natural to inquire how this criterion measure behaves for spectrum estimators. Lomnicki and Zaremba [60] have shown the properties a weighting function must have in order for the integrated mean-square error of its corresponding spectrum estimator to go to zero as the sample record length increases. Both the Tukey-Hanning and Parzen weighting functions have these properties. They have also deduced the form of the weighting function that minimizes the error for a given T, but this form depends on the unknown parameters of the underlying sequence.

The value of the integrated mean-square error as a measure of desirability cannot be accepted uncritically. The measure is an unweighted linear combination of the errors at all frequencies, and so all errors are equally important. It may be that more precision is required in one frequency interval than in another, making errors in the first interval more critical than those in the second. Here a weighted integrated mean-square error would be more meaningful.

An appropriate criterion for choosing among averaging kernels remains to be developed.† For now, we consider a kernel acceptable

---

† For a discussion of other figures of merit, see [48, p. 161 and 73, pp. 185–188].

if it has small side lobes to guard against leakage and permits a reasonable compromise between resolution and stability.† The Tukey-Hanning and Parzen averaging kernels have come into common use because they have these properties, the practical importance of which should not be underestimated.

3.9. The Distribution of Spectrum Estimates

The derivation of the probability law governing spectrum estimates for a discrete normal process may be found in Grenander and Szegö [33]. Owing to its rather complicated form, it has become commonplace to approximate the probability distribution by that of a multiple of *chi*-square, the justification following from certain large sample properties of the classical periodogram.

We begin our description by considering the periodogram properties of the uncorrelated sequence $\{\epsilon_t\}$ defined in Expression (3.2). Consider the spectral representation [5, pp. 274–279]

$$\epsilon(\lambda) = \epsilon_1(\lambda) + i\epsilon_2(\lambda) = (1/T)^{\frac{1}{2}} \sum_{t=1}^{T} \epsilon_t e^{-i\lambda t}$$

that leads to the periodogram

$$I_\epsilon(\lambda) = |\epsilon(\lambda)|^2 = \epsilon_1^2(\lambda) + \epsilon_2^2(\lambda) = (1/T)\left|\sum_{t=1}^{T} \epsilon_t e^{-i\lambda t}\right|^2.$$

Expanding and rearranging the terms on the right leads to

$$I_\epsilon(\lambda) = \sum_{\tau=-T+1}^{T-1} C_{\epsilon,\tau} e^{-i\lambda\tau}$$

$$C_{\epsilon,\tau} = T^{-1} \sum_{t=1}^{T-|\tau|} \epsilon_t \epsilon_{t+\tau}.$$

To determine the statistical properties of $\epsilon_1(\lambda)$ and $\epsilon_2(\lambda)$, we write

$$E[\epsilon^2(\lambda)] = E[\epsilon_1^2(\lambda)] - E[\epsilon_2^2(\lambda)] + 2iE[\epsilon_1(\lambda)\epsilon_2(\lambda)]$$

$$= E\left[(1/T) \sum_{s,t=1}^{T} \epsilon_s \epsilon_t e^{-i\lambda(s+t)}\right]$$

$$= e^{-i\lambda(T+1)} \sin \lambda T/(T \sin \lambda),$$

---

† A nonnegative kernel is also beneficial, but not crucial.

and

$$E[|\epsilon(\lambda)|^2] = E[\epsilon_1^2(\lambda)] + E[\epsilon_2^2(\lambda)]$$

$$= E\left[(1/T) \sum_{s,t=1}^{T} \epsilon_s \epsilon_t e^{-i\lambda(s-t)}\right]$$

$$= 1.$$

Let $\lambda_p = 2\pi p/T$, $\lambda_q = 2\pi q/T$; $p, q = 1, 2, \ldots, [T/2]$. Then

$$E[\epsilon_1^2(\lambda_p)] = E[\epsilon_2^2(\lambda_p)] = \tfrac{1}{2},$$

$$E[\epsilon_1(\lambda_p)\epsilon_2(\lambda_p)] = 0,$$

$$E[\epsilon_1^2(0)] = 1, \qquad E[\epsilon_2^2(0)] = 0,$$

$$E[\epsilon_1^2(\pm\pi)] = 1, \qquad E[\epsilon_2^2(\pm\pi)] = 0,$$

so that

$$E[I_\epsilon(\lambda_p)] = 1.$$

In addition

(3.19) $\text{cov}\,[I_\epsilon(\lambda), I_\epsilon(\lambda')]$

$$= \sum_{\tau,\nu=-T+1}^{T-1} \text{cov}\,(C_{\epsilon,\tau}, C_{\epsilon,\nu}) e^{-i(\lambda\tau + \lambda'\nu)}$$

$$= \frac{\varkappa}{T} + \frac{1}{T^2} \left\{ \frac{\sin^2\,[(\lambda + \lambda')T/2]}{\sin^2\,(\lambda + \lambda')} + \frac{\sin^2\,[(\lambda - \lambda')T/2]}{\sin^2\,(\lambda - \lambda')} \right\},$$

using Expression (3.18) which holds exactly. Then

$$\text{var}\,[I_\epsilon(\lambda)] = [1 + \varkappa/T + \sin^2(\lambda T)/(T^2 \sin^2 \lambda)] \qquad \lambda \neq 0$$

$$= [2 + \varkappa/T] \qquad \lambda = 0$$

$$\text{cov}\,[I_\epsilon(\lambda_p), I_\epsilon(\lambda_q)] = \varkappa/T \qquad p \neq q.$$

Suppose that $\{\epsilon_t\}$ is a normal process. Then $\epsilon_1^2(\lambda_p)$ and $\epsilon_2^2(\lambda_p)$ are each distributed as chi-square with one degree of freedom so that $I_\epsilon(\lambda_p)$ is distributed as chi-square with two degrees of freedom. In addition, $I_\epsilon(\lambda_p)$ and $I_\epsilon(\lambda_q)$ with $p \neq q$ are independent.

Now consider the moving average representation as defined by

Expression (3.1). We have

$$X(\lambda) = (1/T)^{\frac{1}{2}} \sum_{t=1}^{T} X_t e^{-i\lambda t}$$

$$I_x(\lambda) = |X(\lambda)|^2 = (1/T)\left|\sum_{t=1}^{T} X_t e^{-i\lambda t}\right|^2$$

$$= \sum_{\tau=-T+1}^{T-1} C_{x,\tau} e^{-i\lambda\tau}.$$

Direct substitution leads to

(3.20) $$X(\lambda) = (1/T)^{\frac{1}{2}} \sum_{t=1}^{T} \sum_{s=0}^{\infty} a_s \epsilon_{t-s} e^{-i\lambda\tau}$$

$$= (1/T)^{\frac{1}{2}} \sum_{s=0}^{\infty} a_s e^{-i\lambda s} \sum_{t=1-s}^{T-s} \epsilon_t e^{-i\lambda t}$$

$$= (1/T)^{\frac{1}{2}} \sum_{s=0}^{\infty} \left[a_s e^{-i\lambda s}\left(\sum_{t=1}^{T} \epsilon_t e^{-i\lambda t} + \sum_{t=1-s}^{\min\binom{T-s}{0}} \epsilon_t e^{-i\lambda t}\right. \right.$$

$$\left. \left. - \sum_{t=\max\binom{T-s+1}{1}}^{T} \epsilon_t e^{-i\lambda t}\right)\right]$$

$$= \left(\sum_{s=0}^{\infty} a_s e^{-i\lambda s}\right)[\epsilon(\lambda) + 0(T^{-\frac{1}{2}})],$$

provided that a_s goes to zero sufficiently fast. Then

(3.21) $$I_x(\lambda) = 2\pi g(\lambda)[I_\epsilon(\lambda) + 0(1/T)],$$

since

$$g(\lambda) = (2\pi)^{-1}\left|\sum_{s=0}^{\infty} a_s e^{-i\lambda s}\right|^2.$$

It is clear that, apart from the multiplication by $2\pi g(\lambda)$, $I_x(\lambda)$ asymptotically has the same sampling properties as $I_\epsilon(\lambda)$. This fact leads to drawbacks as well as benefits. While $I_x(\lambda)$ is asymptotically unbiased, it is not a consistent estimator of $g(\lambda)$ since the variance of $I_\epsilon(\lambda)$ is invariant with respect to T. On the beneficial side $\{I_x(\lambda_p),$ $\lambda_p = 2\pi p/T; p = \pm1, \pm2, \ldots, \pm[T/2]\}$ is a sequence of independent variates (except for $p = -p$) each proportional to chi-square.

Consider the summation

$$(2\pi/T) \sum_{p=-[T/2]}^{[T/2]} K_M(\lambda - \lambda_p)I_x(\lambda_p)$$

$$= (2\pi/T) \sum_{p=-[T/2]}^{[T/2]} K_M(\lambda - \lambda_p) \sum_{\tau=-T+1}^{T-1} C_{x,\tau} e^{-i\lambda_p\tau}$$

$$I_x(\lambda_{-p}) = I_x(\lambda_p).$$

As T becomes large the independent ordinates in the summation become denser. We may then replace summation by integration so that

$$\int_{-\pi}^{\pi} K_M(\lambda - \omega)I_x(\omega) \, d\omega = \sum_{\tau=-T+1}^{T-1} \left[\int_{-\pi}^{\pi} K_M(\lambda - \omega)e^{i\omega\tau} \, d\omega \right] C_{x,\tau}$$

$$= \sum_{\tau=-M}^{M} k_{M,\tau} C_{x,\tau} e^{-i\lambda\tau}$$

$$= 2\pi\hat{g}(\lambda),$$

using Expression (3.3). Notice that averaging over the periodogram is equivalent to weighting the sample autocovariance function, our initial approach.

For large T we may neglect the terms $0(1/T)$ in Expression (3.21). If M is sufficiently large so that the spectrum g is relatively constant over the bandwidth of the averaging kernel K, we may write

$$2\pi\hat{g}(\lambda) \sim 2\pi g(\lambda) \int_{-\pi}^{\pi} K_M(\lambda - \omega)I_\epsilon(\omega) \, d\omega.$$

The integral on the right is essentially a weighted summation of chi-square variates when $\{\epsilon_t\}$ is a normal process. The integral is, however, not a chi-square variate. Following Welch [99] we approximate the distribution of the integral by that of chi-square.

Noting that the number of degrees of freedom of a chi-square variate is twice the ratio of its squared mean and its variance, we define the stability of $\hat{g}(\lambda)$ as its number of equivalent degrees of freedom:

(3.22)

$$\text{EDF} = \frac{2E^2[\hat{g}(\lambda)]}{\text{var}\,[\hat{g}(\lambda)]} \sim 2T \left[2\pi \int_{-\pi}^{\pi} K_M^2(\lambda - \omega) \, d\omega \right]^{-1} = 2(\delta_\lambda \Psi_{T,M})^{-1}.$$

The EDF for the three averaging kernels are given in Table 3.2. It follows that, as $\Psi_{T,M}^{-1}$ increases, the random variable

$$w = \Psi_{T,M}^{-\frac{1}{2}} \left[\frac{\hat{g}(\lambda)}{g(\lambda)} - 1 \right]$$

asymptotically has a normal distribution with zero mean and unit variance.

If X is not normal, then, as T increases, the distribution of w still converges to the normal distribution with mean zero but with variance $(1 + \varkappa c/M)$, c being defined on p. 98.†

3.10. The Fast Fourier Transform

Section 3.9 showed that computing spectrum estimates by averaging over the periodogram is equivalent to computing them by taking the Fourier transform of the weighted sample autocovariance function. When it was realized that the periodogram ordinates were not consistent estimators, Daniell suggested averaging over the periodogram as a way of improving stability [5, p. 280]. To follow his suggestion required computing all $[T/2] + 1$ periodogram ordinates, a task generally infeasible without the use of a high-speed computer. Even with such assistance, the approach was impractical because of the time-consuming calculations for even moderately long time series.

Emphasis therefore shifted to weighting the sample autocovariance function in the time domain where only M sample autocovariances had to be computed, a task requiring considerably less computing time. This approach has remained the commonly employed one, but recently rediscovered computing techniques are currently shifting emphasis back to periodogram averaging. In 1965, Cooley and Tukey [13] described an algorithm for computing the Fourier coefficient $X(\lambda)$ with a considerable saving in computing time over more direct methods. The algorithm is called the *fast Fourier transform* (FFT) technique. It reduces computing time for Fourier coefficients from order T^2 to order $T \log_2 T$. For example, a time series of length $T = 200$ requires computing time of order 40,000 with direct methods and of order 1,529 with the FFT. Since the computing time saving is especially significant for large T and multivariate spectrum analyses, periodogram averaging will no doubt become the preferred method of spectrum estimation.

---

† See Lomnicki and Zaremba [61, pp. 130–133].

Since the appearance of the Cooley and Tukey paper, Rudnick [82] has discovered a paper by Danielson and Lanczos [17] which described the FFT technique in 1924. More recently, Tukey [91] has noted that a number of earlier papers describing the method have been found, the Danielson and Lanczos one being the oldest.

The FFT technique is a computational procedure that need not concern us here.† Questions about resolution and stability, however, deserve attention. Since a comprehensive statistical literature on periodogram averaging is yet to appear, the present account should be regarded more as a progress report than as a definitive description.

Since the FFT enables one to work directly with the Fourier coefficients, emphasis concerning resolution has recently shifted from the spectrum to the coefficients themselves.‡ In general we may rapidly compute the *modified* Fourier coefficients

$$\tilde{X}(\lambda_p) = \sum_{t=1}^{T} X_t w_{T,t} e^{-i\lambda_p t}$$

$$\lambda_p = 2\pi p/T, \qquad p = 0, \pm 1, \ldots, \pm n = [T/2],$$

where $\{X_t\}$ is the linear process defined in Section 3.1, and the *time series weighting function* w is defined as

$$w_{T,t} = w_{T,T-t+1} \qquad t = 1, 2, \ldots, T,$$

$$W_T(\lambda) = (2\pi)^{-\frac{1}{2}} \sum_{t=1}^{T} w_{T,t} e^{-i\lambda t},$$

$$\int_{-\pi}^{\pi} |W_T(\lambda)|^2 \, d\lambda = \sum_{t=1}^{T} w_{T,t}^2 = 1.$$

The *modified* periodogram ordinate

$$\tilde{I}(\lambda_p) = |\tilde{X}(\lambda_p)|^2$$

has expectation

$$E[\tilde{I}(\lambda_p)] = 2\pi \int_{-\pi}^{\pi} |W_T(\lambda_p - \omega)|^2 g(\omega) \, d\omega.$$

Using the identity

$$E(\theta_1\theta_2\theta_3\theta_4) = E(\theta_1\theta_2)E(\theta_3\theta_4) + E(\theta_1\theta_3)E(\theta_2\theta_4) + E(\theta_1\theta_4)E(\theta_2\theta_3)$$

----

† See [10] for details.
‡ Welch [99] describes a slightly different approach to computing modified Fourier coefficients.

for normal variates θ_1, θ_2, θ_3 and θ_4, with zero means, we have, when $\{X_t\}$ is a normal process,

$$E[\tilde{I}(\lambda_p)\tilde{I}(\lambda_q)] = E|\tilde{X}(\lambda_p)|^2 E|\tilde{X}(\lambda_q)|^2 + E[\tilde{X}(\lambda_p)\tilde{X}^*(\lambda_q)]E[\tilde{X}^*(\lambda_p)\tilde{X}(\lambda_q)]$$
$$+ E[\tilde{X}(\lambda_p)\tilde{X}(\lambda_q)]E[\tilde{X}^*(\lambda_p)\tilde{X}^*(\lambda_q)],$$

$$\text{cov}\,[\tilde{I}(\lambda_p),\,\tilde{I}(\lambda_q)] = \left|2\pi \int_{-\pi}^{\pi} W_T(\lambda_p - \omega)W_T^*(\lambda_q - \omega)g(\omega)\,d\omega\right|^2$$
$$+ \left|2\pi \int_{-\pi}^{\pi} W_T(\lambda_p - \omega)W_T(\lambda_q + \omega)g(\omega)\,d\omega\right|^2.$$

If the spectrum g changes slowly over the interval (λ_p, λ_q), then

$$g(\lambda_p) \sim g(\lambda_q)$$
$$\text{cov}\,[\tilde{I}(\lambda_p),\,\tilde{I}(\lambda_q)] \sim [2\pi g(\lambda_p)]^2[V_T(\lambda_p - \lambda_q) + V_T(\lambda_p + \lambda_q)]$$
$$V_T(\lambda) = \left|\sum_{t=1}^{T} w_{T,t}^2 e^{-i\lambda t}\right|^2.$$

The *uniform* time series weighting function is

$$w_{T,t} = 1/T^{\frac{1}{2}},$$

with corresponding averaging kernel

$$|W_T(\lambda)|^2 = \sin^2{(\lambda T/2)}/[2\pi T \sin^2{(\lambda/2)}],$$

and covariance weights

$$V_T(\lambda_p) = \begin{cases} 1 & \lambda_p = 0,\ \pm\pi \\ 0 & \lambda_p \neq 0,\ \pm\pi. \end{cases}$$

The resulting ordinates form the *raw periodogram*. This choice of weighting function corresponds to a unit weighting of the biased sample autocovariance function as shown in Section 3.9. Recognition of the divisor bias accounts for $|W_T(\lambda)|^2$ assuming the above form instead of the form in Expression (3.13) which follows from a unit weighting of the true autocovariance function.

To improve resolution, the cosine bell weighting function

$$w_{T,t} = [8/(3T)]^{\frac{1}{2}} \sin^2{(\pi t/T)}$$
$$= [2/(3T)]^{\frac{1}{2}}[1 - \cos{(2\pi t/T)}]$$

with averaging kernel

$$|W_T(\lambda)|^2 = \frac{\sin^2{(\lambda T/2)}}{12\pi T}\left\{\frac{\sin\lambda}{\sin^2{(\lambda/2)}}\right.$$
$$\left. - \frac{1}{2}\left[\frac{\sin{(\lambda - 2\pi/T)}}{\sin^2{\frac{1}{2}(\lambda - 2\pi/T)}} + \frac{\sin{(\lambda + 2\pi/T)}}{\sin^2{\frac{1}{2}(\lambda + 2\pi/T)}}\right]\right\}^2$$

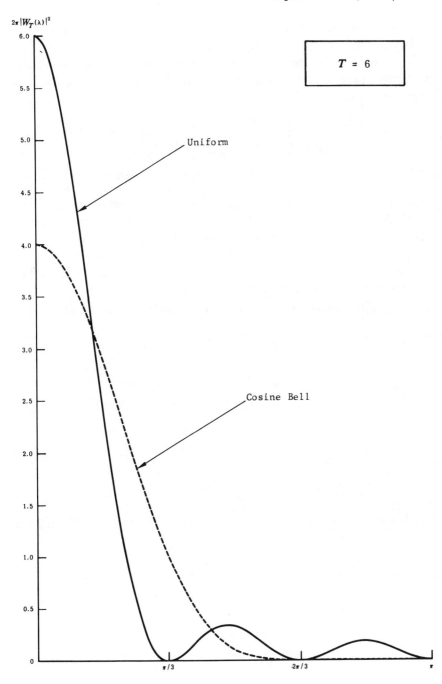

Fig. 20 Averaging kernels corresponding to time series weighting functions

has been suggested in [87]. In Figure 20 the reduction in side lobes indicates the improved resolving power.

Unfortunately, the improved resolution is gained at the cost of inducing positive correlation between the *modified periodogram* ordinates as indicated by †

$$
V_T(\lambda_p) =
\begin{cases}
1 & p = 0, \pm n \\
\frac{4}{9} & p = \pm 1, \pm(n-1) \\
\frac{1}{36} & p = \pm 2, \pm(n-2) \\
0 & p = \pm 3, \pm 4, \ldots, \pm(n-3).
\end{cases}
$$

Since the corresponding spectrum estimate is an average of positively correlated modified periodogram ordinates, it has a larger variance than that obtained using raw periodogram ordinates. Bingham et al. [10] have suggested a modification of the cosine bell weighting function designed to improve the balance between resolution and stability but, as they note, definitive work remains to be done.

To gain stability, we average over the periodogram so that

$$
\hat{g}(\lambda_p) = (1/2\pi) \sum_{q=-m}^{m} K_{m,q} \tilde{I}(\lambda_{p+q})
$$

$$
K_{m,q} = K_{m,-q}
$$

$$
K_{m,q} = 0, \qquad |q| > m
$$

$$
\sum_{q=-m}^{m} K_{m,q} = 1
$$

$$
k_{m,t} = \sum_{q=-m}^{m} K_{m,q} e^{i2\pi q t/T}.
$$

Each spectrum estimate is a weighted sum of $2m+1$ ordinates and has expectation

$$
E[\hat{g}(\lambda_p)] = \int_{-\pi}^{\pi} H_m(\lambda_p - \omega) g(\omega)\, d\omega
$$

$$
H_m(\lambda) = (1/2\pi) \sum_{q=-m}^{m} K_{m,q} |W_T(\lambda_q + \lambda)|^2
$$

$$
= (1/2\pi) \sum_{s,t=1}^{T} w_{T,s} w_{T,t} k_{m,s-t} e^{-i\lambda(s-t)},
$$

---

† Here we assume T to be even.

and asymptotic covariance function

$$\text{cov}\,[\hat{g}(\lambda_p),\,g(\lambda_q)] \sim g(\lambda_p)g(\lambda_q) \sum_{r,s=-m}^{m} K_{m,r}K_{m,s}[V_T(\lambda_{p+r} - \lambda_{q+s})$$

$$+ V_T(\lambda_{p+r} + \lambda_{q+s})]$$

$$= g(\lambda_p)g(\lambda_q) \sum_{s,t=1}^{T} [(w_{T,s}w_{T,t}k_{m,s-t})^2$$

$$\times (e^{-i(\lambda_p-\lambda_q)(s-t)} + e^{-i(\lambda_p+\lambda_q)(s-t)})].$$

When the sample covariance approach to spectrum estimation is used, the emphasis is on finding a time-domain weighting function whose corresponding averaging kernel has small side lobes. In the periodogram averaging approach, we can be more direct and choose an averaging kernel that has no side lobes and concentrates all its area over a specified frequency interval. By doing so, our control over leakage is considerably enhanced.

The rectangular averaging kernel

$$K_{m,q} = \begin{cases} 1/(2m+1) & q = 0,\,\pm 1,\,\pm 2,\,\ldots,\,\pm m \\ 0 & |q| > m \end{cases}$$

is the simplest suggestion and leads to a time-domain weighting function

$$k_{m,t} = \sin\,[2\pi(m+1/2)t/T]/[(2m+1)\sin\,(2\pi t/T)].$$

Using raw periodogram ordinates, the asymptotic variance is

$$\text{var}\,[\hat{g}(\lambda_r)]/[\delta_{\lambda_r}g^2(\lambda_r)] \sim 1/(2m+1),$$

where

$$\lambda_r = 2\pi r/T, \qquad r = 0,\,(2m+1),\,2(2m+1),\,\ldots,\,[T/(4m+2)].$$

Here, spectrum estimates are computed over nonoverlapping intervals and consequently are independent. For moderate T, the $[T/(4m+2)] + 1$ estimates may not provide an adequate graph of the sample spectrum. For overlapping intervals, the asymptotic variance continues to hold when

$$r = 0,\,m+1,\,m+2,\,\ldots,\,n-m-2,\,n-m-1,\,n,$$

but the resulting estimates are correlated. For $0 < r \leq m$ and $n - m \leq r < n$, the asymptotic variance is between the limits given because of the perfect correlation among some of the periodogram ordinates in the average.

Modified ordinates using the cosine bell time series weighting function yield

$$\text{var}\,[\hat{g}(\lambda_r)]/[\delta_{\lambda_r}g^2(\lambda_r)] \sim 2/(2m+1),$$

the increase in variance being due to the positive correlation among ordinates induced by w. In this case nonoverlapping estimates may also be correlated for the same reason.

To improve resolution one may choose K as

$$K_{m,q} = \begin{cases} (m+1-|q|)/(m+1)^2 & |q| \le m \\ 0 & |q| > m. \end{cases}$$

With raw periodogram ordinates, the asymptotic variance is

$$\text{var}\,[\hat{g}(\lambda_r)]/[\delta_{\lambda_r}g^2(\lambda_r)] \sim 2/[3(m+1)];$$

with modified periodogram ordinates using the cosine bell weighting function, it is

$$\text{var}\,[\hat{g}(\lambda_r)]/[\delta_{\lambda_r}g^2(\lambda_r)] \sim 4/[3(m+1)].$$

The computation of equivalent degrees of freedom is direct.

Many problems remain to be discussed in addition to choosing a time series weighting function w and a periodogram averaging kernel K. Among them are the use of the sample mean $\overline{X}$ for the population mean μ, the application of the FFT technique to cross-spectrum estimation, and the choice of an appropriate m in periodogram averaging.

The remarks made here about periodogram estimation and averaging are to be regarded as preliminary, for many subtleties that we have cursorily treated will be more thoroughly examined as periodogram averaging becomes more common. In the remainder of this study we shall concentrate on the sample autocovariance approach to spectrum estimation.

Once again we emphasize that the periodogram averaging approach and the sample autocovariance approach to spectrum estimation share the same statistical theory and differ only in computational method.

3.11. Prewhitening

In general the spectrum varies by orders of magnitude over the frequency interval $(-\pi, \pi)$. The concentration of variance in particular frequency bands is a consequence of the kind of autocorrelation in the sequence. While it is our principal purpose to determine the magnitudes and locations of these concentrations, their presence causes a significant loss of efficiency in the estimation procedure unless they are reduced in magnitude beforehand. Spectrum estimation involves averaging over the periodogram. The flatter the periodogram, the fewer the number of lags needed to resolve the spectrum adequately and, therefore, the higher the efficiency of the resulting estimates. In most spectra there is a strong concentration of variance at the low frequencies. Since we are averaging, it may occur that low-frequency components contribute significantly to estimates in the high-frequency bands because too small a number of lags is used. Similar effects could occur because of the seasonal peaks. In general the low-frequency concentration is several orders of magnitude greater than the magnitudes of the seasonal peaks so that the principal concern is reducing low-frequency leakage.

It therefore becomes desirable to flatten the spectrum as much as possible in order to conserve on lags. It is equally important, however, to recover the estimated true shape of the spectrum after the estimation is performed. To flatten and then recover the spectrum we use our knowledge of linear filtering. Techniques for flattening the spectrum come under the heading of *prewhitening*.

Suppose we generate a new sequence Y by a linear transformation of X so that

$$Y_t = \sum_{s=-p}^{q} a_s X_{t-s}.$$

The corresponding spectrum of Y is

$$g_y(\lambda) = |A(\lambda)|^2 g_x(\lambda),$$

$$A(\lambda) = \sum_{s=-p}^{q} a_s e^{-i\lambda s},$$

so that we have the spectrum of X as

$$g_x(\lambda) = g_y(\lambda)/|A(\lambda)|^2.$$

As an example, let

$$a_s = \begin{cases} 1 & s = 0 \quad\quad 0 < \alpha < 1, \\ \alpha & s = 1 \\ 0 & \text{elsewhere} \end{cases}$$

so that

$$Y_t = X_t - \alpha X_{t-1},$$

$$g_y(\lambda) = (1 - 2\alpha \cos \omega + \alpha^2) g_x(\lambda).$$

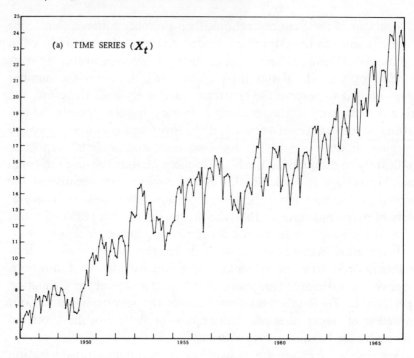

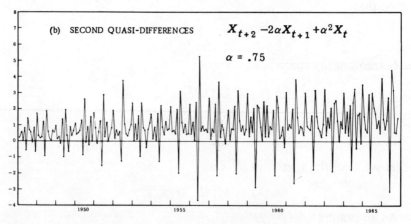

Fig. 21 Manufacturers' shipments of durable goods, seasonally unadjusted, monthly 1947–1966 in billions of dollars

The effect of this linear filtering is to attenuate the low-frequency content of X and to amplify its high-frequency content. The spectrum of Y will therefore be flatter than that of X if the latter has a low-frequency concentration. Since the spectrum g_y is flatter than the spectrum g_x, it will vary less over a given bandwidth and will therefore require fewer lags to accomplish good resolution. Once we estimate g_y, we may recover or *recolor* our estimate of g_x by noting that

$$g_x(\lambda) = g_y(\lambda)/(1 - 2\alpha \cos \lambda + \alpha^2).$$

We may extend this *quasi-differencing* procedure to n differences so that

$$Y_t = \sum_{k=0}^{n} (-\alpha)^k \binom{n}{k} X_{t-k},$$

$$g_y(\lambda) = (1 - 2\alpha \cos \lambda + \alpha^2)^n g_x(\lambda).$$

By a judicious choice of α and n it is possible to flatten the spectrum in the low-frequency range. Upon estimating $g_y(\lambda)$, it is then possible to *recolor* (recover) the estimate of $g_x(\lambda)$ by dividing through by $(1 - 2\alpha \cos \lambda + \alpha^2)^n$. This particular form of prewhitening is especially convenient since the frequency response function only depends on α and n and no additional terms need be added as the number of lags is increased in the time-domain filtering. One finds an α of about 0.75 and an n of 2 or 3 to be satisfactory in a large number of cases.†

Figure 21 shows the original and the second quasi-differenced ($\alpha = 0.75$) time series of shipments. The prewhitened series shows no pronounced trend as the original series does. Figure 22 shows the corresponding recolored and prewhitened sample spectra. Notice that prewhitening has significantly reduced the low-frequency content by two orders of magnitude. Notice also that the first seasonal peak in the prewhitened sample spectra is much smaller than the remaining ones, suggesting that the 12-month cycle is not unusually strong. In fact, recoloring obscures the 12-month peak almost entirely. For the particular α used, prewhitening attenuates the spectrum below

$$\lambda_0 = \cos^{-1}(\alpha/2) = .38\pi$$

and amplifies it for greater frequencies. Figure 22 bears this out.

In suggesting that spectrum estimates be recolored by the appropriate division, we mean to imply there is an advantage in including low-frequency variance concentrations in the spectrum. It would be wasteful to perform spectrum analysis on an economic time series without accounting for such components, since the study of intertemporal development of a series over long periods is fundamental to our understanding of how the underlying process evolves.

One may also use autoregressive techniques to prewhiten a time

---

† See Nerlove [68, pp. 257–258].

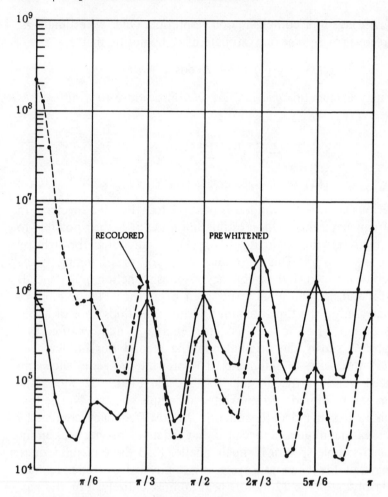

Fig. 22 Sample spectrum of manufacturers' shipments of durable goods, seasonally unadjusted

series [62, 76]. Here an autoregressive scheme is fit to the time series in a manner designed to retain only the most significant coefficients. The autoregressive scheme may then be used to prewhiten the time series. The frequency response function containing the estimated coefficients can then be used to recolor the sample spectrum.

As we discussed in Section 2.23, removing a polynomial in t of order p is equivalent to working with the pth differences of the original time series. We have seen that the effect of this filter is to attenuate the spectrum sharply at low frequencies. For this reason, we consider the quasi-differencing procedure advantageous because it permits a more judicious flattening of the spectrum than the variate-difference method does. If it were our purpose to work with trend-free data, then a least-square polynomial regression would be desirable, for it is known to be asymptotically efficient.† Since our purpose is to remove trend before estimation and then recolor the estimates, we are inclined to recommend prewhitening as described here.

In addition to trend, economic time series usually contain seasonal components that cause peaks in the spectrum at seasonal frequencies. These peaks will be sources of leakage but their importance may be reduced by the transformation

$$Y_t = X_t - \beta X_{t-12},$$

which yields a spectrum

$$g_y(\lambda) = (1 - 2\beta \cos 12\lambda + \beta^2)g_x(\lambda).$$

Observe that this prewhitening procedure attenuates all seasonal frequencies, as well as the low-frequency content of X. Should this attenuation be insufficient to flatten the low-frequency content, a further transformation is possible so that, for example,

$$Z_t = Y_t - \alpha Y_{t-1} = X_t - \alpha X_{t-1} - \beta(X_{t-12} - \alpha X_{t-13}),$$

$$g_z(\lambda) = (1 - 2\alpha \cos \lambda + \alpha^2)g_y(\lambda)$$

$$= (1 - 2\alpha \cos \lambda + \alpha^2)(1 - 2\beta \cos 12\lambda + \beta^2)g_x(\lambda).$$

As an alternative to removing the seasonal peaks by quasi-differencing, one may remove them by means of a harmonic regression. This procedure has asymptotically desirable properties.‡ The seasonal component in economic phenomenon is usually changing character so that one must view a straightforward harmonic regression with skepticism. Hannan [40] has suggested a more complex harmonic regression that accounts for the seasonally changing pattern. Again, since our intention is to recolor the spectrum estimates, we consider it advisable to rely on the quasi-differencing procedure.

The extent to which prewhitening flattens the spectrum depends on the particular time series being studied. If the spectrum is relatively flat, then little if any prewhitening is necessary. Since we are studying time series with a finite number of observations, it is also important to

---

† See Hannan [36, pp. 122–128].
‡ Ibid., p. 127.

note that taking first quasi-differences between successive observations reduces this number by one. Taking the quasi-differences between observations a year apart, however, reduces the number by twelve. In this latter case, one must decide if the improved resolution is worth the loss of these observations. This is partially determined by whether or not the ratio M/T is appreciably reduced after prewhitening.

The purpose of prewhitening is to derive well-resolved estimates of the spectrum while economizing on the number of lags M used in the estimation procedure. As we have shown, stability is inversely proportional to M/T. Since economic time series are seldom very long compared to the extent of autocorrelation in them, it is a natural imperative for the econometrician to take pains in prewhitening so as to keep M as small as possible for reasonably good resolution.

It has long been a common belief among engineers that, if a time series were subjected to prewhitening, the resulting series would be more normal than the original one. Rosenblatt [80] has shown this conjecture to be true for processes satisfying certain general conditions. A quick and simple way to check the credibility of the normal assumption is to plot the sample distribution function of the prewhitened observations on probability paper. If the function is reasonably linear, then the evidence supports the normal assumption.†

Figure 23 shows the sample distribution function plotted on normal probability paper. The observations were first adjusted by subtracting the sample mean and dividing by the sample standard deviation. The straight line corresponds to a normal distribution with zero mean and unit variance. Notice the departure of the sample curve from the straight line at large negative deviations. This implies a skewness to the right in the density function. We must, therefore, regard as extremely tenuous any test results that assume this sequence is normal.

3.12. Bivariate Sampling Properties

Since the method of deriving the sampling properties for the covariance function is analogous to that for autocorrelation analysis, we shall merely state them and mention the appropriate analogies. The cross spectrum has certain worthwhile sampling properties that differ from those of the spectrum; we therefore shall discuss them in some detail. The sampling properties of coherence, gain, and phase angle are then described along with several pertinent statistical tests.

---

† For the use of probability paper see, for example, Schmid [84, pp. 156–160].

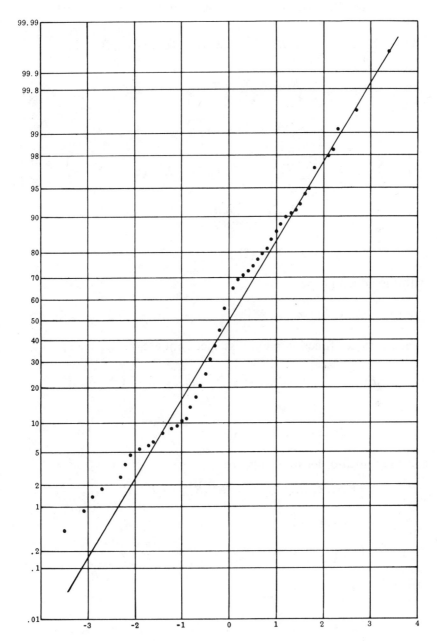

Fig. 23 Sample distribution function of manufacturers' shipments of durable goods, seasonally unadjusted

Consider the discrete normal linear processes X and Y with

$$X_t = \sum_{s=0}^{\infty} a_s \epsilon_{t-s}, \quad \sum_{s=0}^{\infty} a_s^2 < \infty, \quad E(\epsilon_t) = 0, \quad E(\epsilon_t \epsilon_{t+s}) = \begin{cases} 1 & s = 0 \\ 0 & s \neq 0, \end{cases}$$

$$R_{x,\tau} = \sum_{s=0}^{\infty} a_s a_{s+\tau},$$

$$g_x(\lambda) = (2\pi)^{-1} \left| \sum_{s=0}^{\infty} a_s e^{-i\lambda s} \right|^2,$$

$$Y_t = \sum_{s=0}^{\infty} b_s \eta_{t-s}, \quad \sum_{s=0}^{\infty} b_s^2 < \infty, \quad E(\eta_t) = 0, \quad E(\eta_t \eta_{t+s}) = \begin{cases} 1 & s = 0 \\ 0 & s \neq 0, \end{cases}$$

$$R_{y,\tau} = \sum_{s=0}^{\infty} b_s b_{s+\tau},$$

$$g_y(\lambda) = (2\pi)^{-1} \left| \sum_{s=0}^{\infty} b_s e^{-i\lambda s} \right|^2,$$

with the added property

$$E(\epsilon_t \eta_{t+s}) = \begin{cases} \rho & s = \tau^* \\ 0 & s \neq \tau^*, \end{cases} \quad |\rho| < 1.$$

The covariance function and cross spectrum are, respectively,

$$E(X_t Y_{t+\tau}) = R_{xy,\tau} = \rho \sum_{s=0}^{\infty} a_s b_{s+\tau-\tau^*} \leq (R_{x,\tau} R_{y,\tau})^{\frac{1}{2}},$$

$$g_{xy}(\lambda) = (2\pi)^{-1} \rho \left(\sum_{s=0}^{\infty} a_s e^{i\lambda s} \right) \left(\sum_{s=0}^{\infty} b_s e^{-i\lambda s} \right),$$

and the fourth-order moments are

$$(3.23) \quad \begin{cases} E(X_t X_{t+\tau} X_s X_{s+\nu}) = R_{x,\tau} R_{x,\nu} + R_{x,u} R_{x,u+\tau-\nu} \\ \qquad\qquad\qquad\qquad + R_{x,u-\nu} R_{x,u+\tau}, \\ E(Y_t Y_{t+\tau} Y_s Y_{s+\nu}) = R_{y,\tau} R_{y,\nu} + R_{y,u} R_{y,u+\tau-\nu} \\ \qquad\qquad\qquad\qquad + R_{y,u-\nu} R_{y,u+\tau}, \\ E(X_t X_{t+\tau} X_s Y_{s+\nu}) = R_{x,\tau} R_{xy,\nu} + R_{x,u} R_{xy,-u-\tau+\nu} \\ \qquad\qquad\qquad\qquad + R_{xy,-u+\nu} R_{x,u+\tau}, \\ E(Y_t Y_{t+\tau} X_s Y_{s+\nu}) = R_{y,\tau} R_{xy,\nu} + R_{xy,u} R_{y,u+\tau-\nu} \\ \qquad\qquad\qquad\qquad + R_{xy,u+\tau} R_{y,u-\nu}, \\ E(X_t X_{t+\tau} Y_s Y_{s+\nu}) = R_{x,\tau} R_{y,\nu} + R_{xy,-u} R_{xy,-u-\tau+\nu} \\ \qquad\qquad\qquad\qquad + R_{xy,-u+\nu} R_{xy,-u-\tau}, \\ E(X_t Y_{t+\tau} X_s Y_{s+\nu}) = R_{xy,\tau} R_{xy,\nu} + R_{x,u} R_{y,u+\tau-\nu} \\ \qquad\qquad\qquad\qquad + R_{xy,-u+\nu} R_{xy,u+\tau}, \end{cases}$$

where

$$u = t - s.$$

3.13. The Sample Covariance Function

The sample covariance function is

$$(3.24) \qquad C_{xy,\tau} = T^{-1} \sum_{t=1}^{T-|\tau|} (X_t - \bar{X})(Y_{t+\tau} - \bar{Y}),$$

the sample mean $\bar{Y}$ being defined analogously to $\bar{X}$ in Section 3.2. For large $T \gg \tau$, the bias is

$$(3.25) \qquad E(C_{xy,\tau} - R_{xy,\tau}) \sim -2\pi g_{xy}(0)/T.$$

We also have the asymptotic covariance relationships:

$$(3.26) \qquad \begin{bmatrix} \text{cov } (C_{x,\tau};\ C_{x,\nu}) \\ \text{cov } (C_{y,\tau};\ C_{y,\nu}) \\ \text{cov } (C_{x,\tau};\ C_{xy,\nu}) \\ \text{cov } (C_{y,\tau};\ C_{xy,\nu}) \\ \text{cov } (C_{x,\tau};\ C_{y,\nu}) \end{bmatrix} \sim \int_{-\pi}^{\pi} H_{T,\tau,\nu}(\omega) \begin{bmatrix} g_x^2(\omega) \\ g_y^2(\omega) \\ g_x(\omega)g_{xy}(\omega) \\ g_y(\omega)g_{xy}(\omega) \\ |g_{xy}(\omega)|^2 \end{bmatrix} d\omega,$$

$$\text{cov } (C_{xy,\tau};\ C_{xy,\nu})$$

$$\sim 2\pi T^{-1} \int_{-\pi}^{\pi} [g_x(\omega)g_y(\omega)e^{i\omega(\nu-\tau)} + g_{xy}^2(\omega)e^{i\omega(\tau+\nu)}]\, d\omega,$$

$$H_{T,\tau,\nu}(\omega) = [4\pi \cos \omega\tau \cos \omega\nu]/T.$$

These expressions follow by straightforward substitution as in Section 3.2.

3.14. The Sample Cross Spectrum

Our purpose is to estimate the cross spectrum:

$$g_{xy} = c(\lambda) - iq(\lambda) = (2\pi)^{-1} \sum_{\tau=-\infty}^{\infty} R_{xy,\tau} e^{-i\lambda\tau},$$

$$c(\lambda) = (2\pi)^{-1} \sum_{\tau=-\infty}^{\infty} R_{xy,\tau} \cos \lambda\tau,$$

$$q(\lambda) = (2\pi)^{-1} \sum_{\tau=-\infty}^{\infty} R_{xy,\tau} \sin \lambda\tau.$$

The corresponding estimators are

$$(3.27) \quad \hat{g}_{xy}(\lambda) = \hat{c}(\lambda) - i\hat{q}(\lambda) = (2\pi)^{-1} \sum_{\tau=-M}^{M} k_{M,\tau} C_{xy,\tau} e^{-i\lambda\tau},$$

$$\hat{c}(\lambda) = (2\pi)^{-1} \sum_{\tau=-M}^{M} k_{M,\tau} C_{xy,\tau} \cos \lambda\tau,$$

$$\hat{q}(\lambda) = (2\pi)^{-1} \sum_{\tau=-M}^{M} k_{M,\tau} C_{xy,\tau} \sin \lambda\tau.$$

In spectrum estimation the weighting function k is always centered on the zero-lagged sample autocovariance. The practice has carried over into cross-spectrum estimation where it may unfortunately lead to adverse results. This occurs when there is a peak in the covariance function R_{xy} at a nonzero lag. Consider the covariance function

$$R_{xy,\tau} = \begin{cases} \alpha^{\nu-\tau} & -\infty < \tau < \nu \\ \beta^{\tau-\nu} & \nu < \tau < \infty \end{cases} \qquad |\alpha|, |\beta| < 1.$$

The corresponding cross spectrum is then

$$g_{xy}(\lambda) = e^{-i\lambda\nu}[(1 - \alpha e^{i\lambda})^{-1} + (1 - \beta e^{-i\lambda})^{-1} - 1],$$

which oscillates with frequency $2\pi/\nu$. If ν is large the oscillation is rapid and the assumption of a smooth cross spectrum is untenable. If the cross-spectrum estimator is given by Expression (3.27), then we are estimating the cross-spectrum average

$$\left[\int_{-\pi}^{\pi} K_M(\lambda - \omega) g_{xy}(\omega) \, d\omega \right],$$

which requires $M \gg \nu$ for good resolution. It is therefore not necessarily true that the numbers of lags used to resolve the spectra of X and Y are sufficient to resolve the cross spectrum.

Consider the expression

$$\hat{\tilde{g}}_{xy}(\lambda) = (2\pi)^{-1} \sum_{\tau=-M}^{M} k_{M,\tau} C_{xy,\tau+\nu} e^{-i\lambda\tau},$$

which is easily shown to be an estimator of

$$\left[\int_{-\pi}^{\pi} K_M(\lambda - \omega) \tilde{g}_{xy}(\omega) \, d\omega \right]$$

where

$$\tilde{g}_{xy}(\lambda) = e^{i\lambda\nu} g_{xy}(\lambda).$$

The function $\tilde{g}_{xy}$ is not periodic, and, hence, we can apply the same considerations to its estimation that we apply to the spectrum. To recover the cross-spectrum estimator, we use

$$\hat{g}_{xy}(\lambda) = e^{-i\lambda\nu}\hat{\tilde{g}}_{xy}(\lambda).$$

While the covariance function may not be precisely of the form specified above, it is nevertheless desirable to inspect the sample co-variance function for a prominent peak and then center the weighting function k on the peak. This procedure considerably improves the resolution. For convenience of notation we assume ν equals zero and hence Expression (3.27) applies. No loss of generality occurs in the derivation of the sampling properties when the weighting function is shifted and the sample cross spectrum is recovered as described.

As in the univariate case, the averaging procedure and the sample mean are the principal sources of bias so that for large T

$$E[\hat{c}(\lambda)] \sim \int_{-\pi}^{\pi} K_M(\lambda - \omega)c(\omega)\, d\omega - 2\pi c(0)K_M(\lambda)/T,$$

$$E[\hat{q}(\lambda)] \sim \int_{-\pi}^{\pi} K_M(\lambda - \omega)q(\omega)\, d\omega.$$

Notice that the sample mean bias affects the co-spectrum only. Assuming co-spectrum estimates to be well resolved so that

$$\int_{-\pi}^{\pi} K_M(\lambda - \omega)c(\omega)\, d\omega \sim c(\lambda),$$

we may regard

(3.28a) $$\qquad \hat{c}_u(\lambda) = \hat{c}(\lambda) + 2\pi K_M(\lambda)\hat{c}_u(0)/T,$$

(3.28b) $$\qquad \hat{c}_u(0) = \hat{c}(0)[1 - 2\pi K_M(0)/T]^{-1},$$

as an unbiased estimator of the co-spectrum.

While the quadrature-spectrum estimator is not subject to sample mean bias, it suffers from a peculiar bias because $q(\lambda)$ is an odd function. The quadrature-spectrum average is†

$$q'(\lambda) = \int_{-\pi}^{\pi} K_M(\lambda - \omega)q(\omega)\, d\omega$$

$$= \int_{0}^{\pi} K_M(\lambda - \omega)q(\omega)\, d\omega - \int_{0}^{\pi} K_M(\lambda + \omega)q(\omega)\, d\omega,$$

---

† See Hannan [38, p. 36].

since $q(\omega) = -q(-\omega)$. If q is relatively constant over the bandwidth, we have

$$q'(\lambda) \sim q(\lambda) \int_0^\pi [K_M(\lambda - \omega) - K_M(\lambda + \omega)] \, d\omega,$$

where the integral is not necessarily close to or equal to unity. If estimates are made at intervals of π/M, then the estimates at π/M and $(M - 1)\pi/M$ suffer sizable downward bias with both the Tukey-Hanning and the Parzen windows.

It is easily seen that

$$w_M(\lambda) = \int_0^\pi [K_M(\lambda - \omega) - K_M(\lambda + \omega)] \, d\omega$$

$$= (4/\pi) \sum_{\tau=1}^M \{k_{M,\tau}[\sin \lambda\tau \sin^2 (\pi\tau/2)]/\tau\}.$$

To compensate for the bias, we suggest using quadrature-spectrum estimates of the form

(3.28c) $$\hat{q}_u(\lambda) = \hat{q}(\lambda)/w_M(\lambda).$$

It is also to be noted that

(3.28d) $$q(0) = q(\pi) = 0$$

by definition so that it is reasonable to set

$$\hat{q}_u(0) = \hat{q}_u(0) = 0.$$

Hereafter we assume that these bias corrections have been made in both the co- and quadrature-spectrum estimates and we drop the subscript u.

3.15. The Asymptotic Covariances of Cross-Spectrum Estimates†

In a manner analogous to that of Section 3.6, one may show the large sample covariance functions of interest to be

(3.29a)

$$\begin{cases} \operatorname{cov} [\hat{g}_x(\lambda), \hat{g}_x(\lambda')] \\ \operatorname{cov} [\hat{g}_y(\lambda), \hat{g}_y(\lambda')] \\ \operatorname{cov} [\hat{g}_x(\lambda), \hat{g}_{xy}(\lambda')] \\ \operatorname{cov} [\hat{g}_y(\lambda), \hat{g}_{xy}(\lambda')] \\ \operatorname{cov} [\hat{g}_x(\lambda), \hat{g}_y(\lambda')] \end{cases} \sim \int_{-\pi}^\pi J_{T,M}(\omega, \lambda, \lambda') \begin{bmatrix} g_x^2(\omega) \\ g_y^2(\omega) \\ g_x(\omega)g_{xy}(\omega) \\ g_y(\omega)g_{xy}(\omega) \\ |g_{xy}(\omega)|^2 \end{bmatrix} d\omega,$$

---

† See also [49].

(3.29b)

$$\text{cov}\,[\hat{g}_{xy}(\lambda),\,\hat{g}_{xy}(\lambda')] \sim 2\pi T^{-1} \int_{-\pi}^{\pi} [g_x(\omega)g_y(\omega)K_M(\omega + \lambda')$$
$$+ g_{xy}^2(\omega)K_M(\omega - \lambda')]K_M(\omega - \lambda)\,d\omega,$$

(3.29c)

$$\text{cov}\,[\hat{g}_{xy}(\lambda),\,\hat{g}_{xy}^*(\lambda')] \sim 2\pi T^{-1} \int_{-\pi}^{\pi} [g_x(\omega)g_y(\omega)K_M(\omega - \lambda')$$
$$+ g_{xy}^2(\omega)K_M(\omega + \lambda')]K_M(\omega - \lambda)\,d\omega,$$

$$J_{T,M}(\omega, \lambda, \lambda')$$
$$= \pi T^{-1}\{[K_M(\omega - \lambda') + K_M(\omega + \lambda')][K_M(\omega - \lambda) + K_M(\omega + \lambda)]\}.$$

We note that

$$\text{cov}\,[\hat{g}_{xy}(\lambda),\,\hat{g}_{xy}(\lambda')] = \text{cov}\,[\hat{c}(\lambda),\,\hat{c}(\lambda')] - \text{cov}\,[\hat{q}(\lambda),\,\hat{q}(\lambda')]$$
$$- i\{\text{cov}\,[\hat{c}(\lambda),\,\hat{q}(\lambda')] + \text{cov}\,[\hat{q}(\lambda),\,\hat{c}(\lambda')]\},$$

$$\text{cov}\,[\hat{g}_{xy}(\lambda),\,\hat{g}_{xy}^*(\lambda')] = \text{cov}\,[\hat{c}(\lambda),\,\hat{c}(\lambda')] + \text{cov}\,[\hat{q}(\lambda),\,\hat{q}(\lambda')]$$
$$+ i\{\text{cov}\,[\hat{c}(\lambda),\,\hat{q}(\lambda')] - \text{cov}\,[\hat{q}(\lambda),\,\hat{c}(\lambda')]\},$$

$$\hat{c}(\lambda) = \hat{c}(-\lambda), \quad \hat{q}(\lambda) = -\hat{q}(-\lambda),$$

so that

(3.30)
$$\begin{Bmatrix} \text{cov}\,[\hat{c}(\lambda),\,\hat{c}(\lambda')] \\ \text{cov}\,[\hat{q}(\lambda),\,\hat{q}(\lambda')] \\ \text{cov}\,[\hat{c}(\lambda),\,\hat{q}(\lambda')] \end{Bmatrix}$$
$$\sim \int_{-\pi}^{\pi} \begin{bmatrix} J'_{T,M}(\omega, \lambda, \lambda')[g_x(\omega)g_y(\omega) + c^2(\omega) - q^2(\omega)] \\ J''_{T,M}(\omega, \lambda, \lambda')[g_x(\omega)g_y(\omega) + q^2(\omega) - c^2(\omega)] \\ 2J''_{T,M}(\omega, \lambda, \lambda')c(\omega)q(\omega) \end{bmatrix} d\omega,$$

$$J'_{T,M}(\omega, \lambda, \lambda') = \pi[K_M(\omega - \lambda') + K_M(\omega + \lambda')]K_M(\omega - \lambda)/T,$$

$$J''_{T,M}(\omega, \lambda, \lambda') = \pi[K_M(\omega - \lambda') - K_M(\omega + \lambda')]K_M(\omega - \lambda)/T.$$

For well-resolved estimates and $\lambda = \lambda'$ we have the asymptotic covariance matrix

(3.31)

	$\hat{g}_x$	$\hat{g}_y$	$\hat{c}$	$\hat{q}$
$\Psi_{T,M}$	g_x^2	$c^2 + q^2$	$g_x c$	$g_x q$
		g_y^2	$g_y c$	$g_y q$
			$\frac{1}{2}(g_x g_y + c^2 - q^2)$	cq
				$\frac{1}{2}(g_x g_y - c^2 + q^2)$

where the frequency argument $0 < |\lambda| < \pi$ is implicitly assumed. From (3.30) we note that at $|\lambda| = 0, \pi$

$$\text{var}\,[\hat{q}(\lambda)] = 0,$$
$$\text{cov}\,[\hat{c}(\lambda), \hat{q}(\lambda)] = 0,$$

and all the remaining covariances are double those given in (3.13).

3.16. The Distribution of Bivariate Spectrum Estimates

The approximating distribution theory for bivariate spectrum estimates relies heavily on the multivariate distribution theory for real normal random variables as described in Anderson [4] and on the extension of this theory to complex normal random variables as described in Goodman [26, 27]. In fact the purpose of Goodman's work is to provide a theoretical basis for the joint distributions of bivariate spectrum estimates as in [25] and of multivariate spectrum estimates as in [26, 27]. It is instructive to build up to Goodman's bivariate results by first reviewing some well-established results for real normal random variables.

Let

$$\xi = \begin{bmatrix} \xi_1 \\ \xi_2 \end{bmatrix}$$

be a bivariate normal random variable with mean zero and covariance matrix

$$\Sigma = E(\xi\xi') = \begin{bmatrix} \sigma_{11} & \sigma_{12} \\ \sigma_{12} & \sigma_{22} \end{bmatrix},$$

and let

$$\xi_j = \begin{bmatrix} \xi_{1j} \\ \xi_{2j} \end{bmatrix} \qquad j = 1, 2, \ldots, n$$

be a set of n independent observations on ξ. Then the product moment summations

$$\mathbf{A} = \begin{pmatrix} A_{11} & A_{12} \\ A_{21} & A_{22} \end{pmatrix} = \sum_{j=1}^{n} \xi_j\xi_j',$$

$$A_{ik} = \sum_{j=1}^{n} \xi_{ij}\xi_{kj} \qquad i, k = 1, 2$$

jointly have the Wishart distribution with probability density function†

$$\frac{|\mathbf{A}|^{\frac{1}{2}(n-3)} \exp\left[-\frac{1}{2}\,\text{tr}\,(\Sigma^{-1}\mathbf{A})\right]}{2^n \pi^{\frac{1}{2}}|\Sigma|^{n/2}\Gamma(n/2)\Gamma[(n-1)/2]} \qquad n \geq 2,$$

---

† See Anderson [4, p. 154].

which we denote as $w_R(\mathbf{A}|\mathbf{\Sigma}, n)$, and where tr $(\mathbf{\Sigma}^{-1}\mathbf{A})$ denotes the trace of $\mathbf{\Sigma}^{-1}\mathbf{A}$. Also the statistic†

$$|\mathbf{A}|/|\mathbf{\Sigma}| = (A_{11}A_{22} - A_{12}A_{21})/(\sigma_1^2\sigma_2^2 - \sigma_{12}\sigma_{21})$$
$$= [(A_{11} - A_{12}A_{22}^{-1}A_{21})/(\sigma_{11} - \sigma_{12}\sigma_{22}^{-1}\sigma_{21})] \cdot [A_{22}/\sigma_{22}]$$

has the distribution of the product of two independent chi-square variates with $(n - 1)$ and n degrees of freedom respectively.

The sample variance of ξ_2 is

$$s_{22} = A_{22}/n,$$

and the statistic

$$ns_{22}/\sigma_{22} = A_{22}/\sigma_{22}$$

has the chi-square distribution with n degrees of freedom. The population and sample conditional variances of ξ_1 given ξ_2 are

$$\sigma_{11 \cdot 2} = \sigma_{11} - \sigma_{12}\sigma_{22}^{-1}\sigma_{21},$$

$$s_{11 \cdot 2} = (A_{11} - A_{12}A_{22}^{-1}A_{21})/(n - 1),$$

respectively, and the statistic $(n - 1)s_{11 \cdot 2}/\sigma_{11 \cdot 2}$ has the chi-square distribution with $n - 1$ degrees of freedom.

These results provide a background for the study of the sample product moments of two complex normal random variables. Let ξ now be a bivariate complex random variable

$$\xi = \begin{bmatrix} \xi_{11} - i\xi_{12} \\ \xi_{21} - i\xi_{22} \end{bmatrix},$$

where ξ_{11}, ξ_{12}, ξ_{21}, and ξ_{22} are real normal random variables with means zero and covariances

$$E(\xi_{11}^2) = E(\xi_{12}^2) = \sigma_{11}/2,$$
$$E(\xi_{21}^2) = E(\xi_{22}^2) = \sigma_{22}/2,$$
$$E(\xi_{11}\xi_{12}) = E(\xi_{21}\xi_{22}) = 0,$$
$$E(\xi_{11}\xi_{21}) = E(\xi_{12}\xi_{22}) = \alpha\sqrt{\sigma_{11}\sigma_{22}}/2,$$
$$E(\xi_{11}\xi_{22}) = -E(\xi_{12}\xi_{21}) = \beta\sqrt{\sigma_{11}\sigma_{22}}/2,$$
$$0 \le |\alpha|, \ |\beta| < 1; \quad 0 \le \gamma^2 = \alpha^2 + \beta^2 < 1.$$

---

† Ibid., pp. 170–171.

We arrange the covariance in a complex product moment matrix

$$\Sigma = E(\xi^*\xi') = \begin{bmatrix} \sigma_{11} & (\alpha - i\beta)\sigma_{11}\sigma_{22} \\ (\alpha + i\beta)\sigma_{11}\sigma_{22} & \sigma_{22} \end{bmatrix}, \quad |\Sigma| > 0$$

where $*$ and $'$ indicate complex conjugate and transpose respectively.

For a sample of n independent observations on ξ, we have the sample product moment summations

$$\mathbf{A} = \sum_{j=1}^{n} \xi_j^*\xi_j' = \begin{bmatrix} A_{11} & A_{12R} - iA_{12I} \\ A_{21R} + iA_{21I} & A_{22} \end{bmatrix},$$

$$A_{kk} = \sum_{j=1}^{n} (\xi_{k1,j}^2 + \xi_{k2,j}^2) \quad k = 1, 2,$$

$$A_{12R} = A_{21R} = \sum_{j=1}^{n} (\xi_{11,j}\xi_{21,j} + \xi_{12,j}\xi_{22,j}),$$

$$A_{12I} = A_{21I} = \sum_{j=1}^{n} (\xi_{11,j}\xi_{22,j} - \xi_{12,j}\xi_{21,j}).$$

Then, as shown in Goodman [26], A_{11}, A_{22}, A_{12R}, and A_{12I} jointly have the complex Wishart distribution with probability density function

$$\frac{|A|^{n-2} \exp\left[-\text{tr}\,(\Sigma^{-1}A)\right]}{\pi|\Sigma|^n\Gamma(n)\Gamma(n-1)} \quad n \ge 2,$$

which we denote as $w_C(\mathbf{A}|\Sigma, n)$.

The statistic

$$|\mathbf{A}|/|\Sigma| = (A_{11}A_{22} - A_{12R}A_{21R} - A_{12I}A_{21I})/(\sigma_{11}\sigma_{22} - \gamma^2\sigma_{11}\sigma_{22})$$

$$= [A_{11} - A_{12R}A_{22}^{-1}A_{21R} + A_{12I}A_{22}^{-1}A_{21I})]$$

$$/[\sigma_{11}(1 - \gamma^2)] \cdot (A_{22}/\sigma_{22})$$

also has the distribution of the product of two independent chi-square variates with $2(n-1)$ and $2n$ degrees of freedom respectively. The quantity A_{22}/σ_{22} has the chi-square distribution with $2n$ degree of freedom. The population and sample conditional variances of $\xi_{11} - i\xi_{12}$ given $\xi_{21} - i\xi_{22}$ are

$$s_{11\cdot2} = [A_{11} - (A_{12R}A_{22}^{-1}A_{21R} + A_{12I}A_{22}^{-1}A_{21I})]/[2(n-1)],$$

$$\sigma_{11\cdot2} = \sigma_{11}(1 - \gamma^2),$$

respectively, and one may show that $2(n-1)s_{11\cdot2}/\sigma_{11\cdot2}$ has the chi-square distribution with $2(n-1)$ degrees of freedom. These results for a bivariate complex normal random variate bring us a step closer to the approximating distribution for spectrum and cross-spectrum estimates.

To extend these results to the frequency domain, consider the finite Fourier transforms for normal processes X and Y with zero means:

$$X(\lambda) = X_1(\lambda) - iX_2(\lambda) = (1/T)^{\frac{1}{2}} \sum_{t=1}^{T} X_t e^{-i\lambda t},$$

$$Y(\lambda) = Y_1(\lambda) - iY_2(\lambda) = (1/T)^{\frac{1}{2}} \sum_{t=1}^{T} Y_t e^{-i\lambda t}.$$

As in Section 3.9, one may show that asymptotically

$$\left.\begin{matrix} E[X_1(\lambda_p)X_2(\lambda_q)] \\ E[Y_1(\lambda_p)Y_2(\lambda_q)] \end{matrix}\right\} \sim 0,$$

$$\lambda_p = 2\pi p/T, \quad \lambda_q = 2\pi q/T; \quad p, q = 0, 1, \ldots, [T/2],$$

$$\left.\begin{matrix} E[X_1(\lambda_p)Y_1(\lambda_q)] \\ E[X_1(\lambda_p)Y_2(\lambda_q)] \\ E[X_2(\lambda_p)Y_1(\lambda_q)] \\ E[X_2(\lambda_p)Y_2(\lambda_q)] \end{matrix}\right\} \sim 0 \quad p \neq q,$$

and

$$\left.\begin{matrix} E[X_1^2(\lambda_p)] \\ E[X_2^2(\lambda_p)] \end{matrix}\right\} \sim \pi g_x(\lambda_p),$$

$$\left.\begin{matrix} E[Y_1^2(\lambda_p)] \\ E[Y_2^2(\lambda_p)] \end{matrix}\right\} \sim \pi g_y(\lambda_p),$$

$$\left.\begin{matrix} E[X_1(\lambda_p)Y_1(\lambda_p)] \\ E[X_2(\lambda_p)Y_2(\lambda_p)] \end{matrix}\right\} \sim \pi c(\lambda_p),$$

$$\left.\begin{matrix} E[X_1(\lambda_p)Y_2(\lambda_p)] \\ -E[X_2(\lambda_p)Y_1(\lambda_p)] \end{matrix}\right\} \sim \pi q(\lambda_p).$$

We next define the complex vector random variable

$$\xi(\lambda_p) = \begin{bmatrix} X_1(\lambda_p) - iX_2(\lambda_p) \\ Y_1(\lambda_p) - iY_2(\lambda_p) \end{bmatrix},$$

and the matrix periodogram

$$\mathbf{I}(\lambda_p) = \xi^*(\lambda_p)\xi'(\lambda_p)$$

with expectation

$$E[\mathbf{I}(\lambda_p)] \sim 2\pi \mathbf{g}(\lambda_p) = 2\pi \begin{bmatrix} g_x(\lambda_p) & c(\lambda_p) - iq(\lambda_p) \\ c(\lambda_p) + iq(\lambda_p) & g_y(\lambda_p) \end{bmatrix}$$

$$q(0) = q(\pi) = 0.$$

We now form the sample product moment summations

$$2\pi(2m + 1)\tilde{\mathbf{g}}(\lambda_p) = \sum_{q=-m}^{m} \mathbf{I}(\lambda_p + \lambda_q) = \sum_{q=-m}^{m} \xi^*(\lambda_p + \lambda_q)\xi'(\lambda_p + \lambda_q)$$

$$m < p < T/2 - m.$$

Then the statistic $(2m + 1)\tilde{\mathbf{g}}(\lambda_p)$ asymptotically has the complex Wishart distribution with probability density function

$$w_C[(2m + 1)\tilde{\mathbf{g}}(\lambda_p)|\mathbf{g}(\lambda_p), 2m + 1].$$

We also have the estimates

$$2\pi(2m + 1)\tilde{\mathbf{g}}(0) = \sum_{q=-m}^{m} \mathbf{I}(\lambda_q),$$

$$2\pi(2m + 1)\tilde{\mathbf{g}}(\pi) = \sum_{q=-m}^{m} \mathbf{I}(\pi + \lambda_q) \qquad T \text{ even,}$$

$$\mathbf{I}(\lambda_q) = \mathbf{I}^*(\lambda_{-q}), \quad \mathbf{I}(\pi + \lambda_q) = \mathbf{I}^*(\pi + \lambda_{-q}).$$

Then $(2m + 1)\tilde{\mathbf{g}}(0)$ and $(2m + 1)\tilde{\mathbf{g}}(\pi)$ asymptotically have the Wishart distribution with probability density functions

$$w_R[(2m + 1)\tilde{\mathbf{g}}(0)|\mathbf{g}(0), 2m + 1]$$

and

$$w_R[(2m + 1)\tilde{\mathbf{g}}(\pi)|, \mathbf{g}(\pi), 2m + 1]$$

respectively.

To improve resolution we replace $\tilde{\mathbf{g}}$ by $\hat{\mathbf{g}}$ where

$$\hat{\mathbf{g}}(\lambda) = \begin{bmatrix} \hat{g}_x(\lambda) & \hat{c}(\lambda) - i\hat{q}(\lambda) \\ \hat{c}(\lambda) + i\hat{q}(\lambda) & \hat{g}_y(\lambda) \end{bmatrix},$$

as defined in Sections 3.5 and 3.14. Then matching moments as in Section 3.9, we approximate the joint distribution of the elements in $n\hat{\mathbf{g}}(\lambda)$ for $0 < |\lambda| < \pi$ by the complex Wishart distribution with probability density function $w_C[n\hat{\mathbf{g}}(\lambda)|\mathbf{g}(\lambda), n]$, and we approximate the joint distribution of the elements in $n\hat{\mathbf{g}}(\lambda)$ for $|\lambda| = 0, \pi$ by the Wishart distribution with probability density function $w_R[n\hat{g}(\lambda)|g(\lambda), n]$ and where

$$n = 1/\Psi_{T,M}.$$

As a further approximation, we regard the statistic

$$n_\lambda^2|\hat{\mathbf{g}}(\lambda)|/|\mathbf{g}(\lambda)| = n_\lambda^2[\hat{g}_x(\lambda)\hat{g}_y(\lambda) - |\hat{g}_{xy}(\lambda)|^2]/[g_x(\lambda)g_y(\lambda) - |g_{xy}(\lambda)|^2]$$

$$= [n_\lambda\hat{g}_\eta(\lambda)/g_\eta(\lambda)] \cdot [n_\lambda\hat{g}_x(\lambda)/g_x(\lambda)],$$

$$\hat{g}_\eta(\lambda) = \hat{g}_y(\lambda)[1 - \gamma^2(\lambda)],$$

$$g_\eta(\lambda) = g_y(\lambda)[1 - \gamma^2(\lambda)]$$

$$n_\lambda = \varkappa_\lambda/\Psi_{T,M},$$

$$\varkappa_\lambda = 1/\delta_\lambda$$

as the product of two independent chi-square variates with $(n_\lambda - \varkappa_\lambda)$ and n_λ equivalent degrees of freedom respectively. The quantities $\Psi_{T,M}$ and δ_λ are defined on p. 98. Then $n_\lambda\hat{g}_x(\lambda)/g_x(\lambda)$ is treated as a chi-square variate with n_λ equivalent degrees of freedom as before and $n_\lambda\hat{g}_\eta(\lambda)/g_\eta(\lambda)$, as a chi-square variate with $(n_\lambda - \varkappa_\lambda)$ equivalent degrees of freedom.

Using appropriate limiting arguments one may show that, as n_λ increases, the joint distribution of $n_\lambda\hat{g}_x(\lambda)$, $n_\lambda\hat{g}_y(\lambda)$, $n_\lambda\hat{c}(\lambda)$, and $n_\lambda\hat{q}(\lambda)$ converges to the multivariate normal distribution. This result assumes significance in Chap. 4 where it is used to find the asymptotic distribution of estimated coefficients in distributed lag models.

While the approximating nature of the distribution theory cannot be overlooked, the results presented here do serve as valuable guides in evaluating cross-spectrum analyses. This is especially true with regard to the sample residual spectrum $\hat{g}_\eta$ whose appearance as a function of frequency can play a significant role in econometric model building.†

3.17. The Sample Coherence, Gain, and Phase Angle and Their Asymptotic Bias and Variance

The estimators of the coherence, gain, and phase angle are

$$\hat{\gamma}(\lambda) = |\hat{g}_{xy}(\lambda)|/[\hat{g}_x(\lambda)\hat{g}_y(\lambda)]^{\frac{1}{2}},$$

$$\hat{G}(\lambda) = |\hat{g}_{xy}(\lambda)|/\hat{g}_x(\lambda),$$

$$\hat{\phi}(\lambda) = \tan^{-1}[-\hat{q}(\lambda)/\hat{c}(\lambda)]$$

---

† See Sections 2.26 and 2.30.

respectively. To derive bias and variance expressions to order M/T, we assume the sample spectra do not vary much around their respective means and are well resolved. Using the Taylor series expansions we have

$$\hat{\gamma}/\gamma \sim 1 + |g_{xy}|^{-2}[c(\hat{c} - c) + q(\hat{q} - q)]$$
$$- \tfrac{1}{2}[(\hat{g}_x - g_x)/g_x + (\hat{g}_y - g_y)/g_y]$$
$$+ \tfrac{1}{4}[(\hat{g}_x - g_x)(\hat{g}_y - g_y)/(g_x g_y)]$$
$$+ \tfrac{3}{8}[(\hat{g}_x - g_x)^2/g_x^2 + (\hat{g}_y - g_y)^2/g_y^2]$$
$$+ \tfrac{1}{2}|g_{xy}|^{-2}\{[(\hat{c} - c)^2 + (\hat{q} - q)^2] - [c(\hat{c} - c) + q(\hat{q} - q)]$$
$$\times [(\hat{g}_x - g_x)/g_x + (\hat{g}_y - g_y)/g_y]\}$$
$$- \tfrac{1}{2}|g_{xy}|^{-4}[c(\hat{c} - c) + q(\hat{q} - q)]^2;$$

$$\hat{G}/G \sim 1 - (\hat{g}_x - g_x)/g_x + |g_{xy}|^{-2}[c(\hat{c} - c) + q(\hat{q} - q)]$$
$$+ [(\hat{g}_x - g_x)/g_x]^2$$
$$+ \tfrac{1}{2}|g_{xy}|^{-2}\{[(\hat{c} - c)^2 + (\hat{q} - q)^2]$$
$$- 2[c(\hat{c} - c) + q(\hat{q} - q)](\hat{g}_x - g_x)/g_x\}$$
$$- \tfrac{1}{2}|g_{xy}|^{-4}[c(\hat{c} - c) + q(\hat{q} - q)]^2;$$
$$\hat{\phi} \sim \phi + |g_{xy}|^{-2}[q(\hat{c} - c) - c(\hat{q} - q)]$$
$$+ |g_{xy}|^{-4}\{qc[(\hat{q} - q)^2 - (\hat{c} - c)^2]$$
$$+ [(c^2 - q^2)(\hat{c} - c)(\hat{q} - q)]\};$$

so that

$$E(\hat{\gamma} - \gamma) \sim \delta_\lambda \Psi_{T,M}(1 - \gamma^2)^2/(4\gamma),$$
$$E(\hat{G} - G) \sim \delta_\lambda \Psi_{T,M}(1 - \gamma^2)/(4\gamma^2),$$
$$E(\hat{\phi} - \phi) \sim 0,$$
$$\text{var}\,(\hat{\gamma}) \sim \delta_\lambda \Psi_{T,M}(1 - \gamma^2)^2/2,$$
$$\text{var}\,(\hat{G}) \sim \delta_\lambda \Psi_{T,M}(1 - \gamma^2)g_y/(2g_x),$$
$$\text{var}\,(\hat{\phi}) \sim \delta_\lambda \Psi_{T,M}(\gamma^{-2} - 1)/2.$$

Here the frequency argument is implicitly assumed. These results hold for $|g_{xy}|^2 > 0$. In fact, reasonable approximations would require higher order terms in M/T if $|g_{xy}|^2$ were close to zero.

Note that both the bias (except for $\hat{\phi}$) and the variance are inversely related to the coherence γ, implying that if the coherence is low, the estimates are not very reliable.

One may also show that to order M/T,

$$\text{cov}\,(\hat{G}, \hat{\phi}) = 0,$$

so that to this degree of approximation the sample gain and phase angle are uncorrelated.

3.18. The Distributions of Sample Coherence, Gain, and Phase Angle†

By appropriately transforming the approximating joint distribution of $\hat{g}_x(\lambda)$, $\hat{g}_y(\lambda)$, $\hat{c}(\lambda)$, and $\hat{q}(\lambda)$, one may derive the relevant approximating distributions for the sample coherence, gain, and phase angle as shown in Table 8. Here the argument λ is implicit. Amos and Koopmans [3] have prepared tables of the distribution of sample coherence.

It is of particular interest to note that the first two moments of $\hat{\gamma}$ are

$$E(\hat{\gamma}) = (1 - \gamma^2)^n \sum_{j=0}^{\infty} \frac{\Gamma^2(n + j)}{\Gamma^2(j + 1)} B(j + 3/2, n - 1)\gamma^{2j},$$

$$E(\hat{\gamma}^2) = (1 - \gamma^2)^n \sum_{j=0}^{\infty} \frac{\Gamma^2(n + j)}{\Gamma^2(j + 1)} B(j + 2, n - 1)\gamma^{2j},$$

where B is the incomplete beta function:

$$B(p, q) = \int_0^1 t^{p-1}(1 - t)^{q-1}\, dt.$$

Enochson and Goodman [23] have shown that, for n greater than 20 and γ^2 between 0.4 and 0.95, the variate

$$\varphi = \tan h^{-1}(\hat{\gamma})$$

is approximately normally distributed with mean and variance

$$\mu = \tan h^{-1}(\gamma) + \frac{1}{2(n - 1)},$$

and

$$\sigma^2 = \frac{1}{2(n - 1)},$$

respectively.

---

† See Jenkins [49] for an alternative approach to approximating the distributions of sample gain and phase angle.

Table 8

Approximating Distributions of Sample Coherence, Gain, and Phase Angle

z = coherence, g = gain, ϕ = phase angle

Variables	Probability Density Function[a]
z, ϕ	$\dfrac{\delta^n z(1-z^2)^{n-2}}{\pi\Gamma(n)\Gamma(n-1)}\displaystyle\sum_{k=0}^{\infty}\dfrac{2^k\gamma^k\Gamma^2(n+k/2)}{\Gamma(k+1)}z^k\cos^k(\phi-\phi_0)\qquad \delta=1-\gamma^2$
z	$\dfrac{2\delta^n z(1-z^2)^{n-2}}{\Gamma(n)\Gamma(n-1)}\displaystyle\sum_{k=0}^{\infty}\dfrac{\gamma^{2k}\Gamma^2(n+k)}{\Gamma^2(k+1)}z^{2k}$
g, ϕ	$\dfrac{n\delta^n g}{\pi[1-s^2+(g+s)^2]^{n+1}}$
g	$\dfrac{2n\delta^2 g}{\pi(1-2\gamma g+g^2)^{n+\frac12}(1+2\gamma g+g^2)^{\frac12}}\displaystyle\sum_{k=0}^{\infty}\dfrac{\Gamma(k+\frac12)\Gamma(n-k+\frac12)}{\Gamma(k+1)\Gamma(n-k+1)}\left(\dfrac{1-2\gamma g+g^2}{1+2\gamma g+g^2}\right)^k$
ϕ	$\dfrac{\delta^n}{2\pi}-\dfrac{n\delta^n s}{2\pi(1-s^2)^{n+\frac12}}\left[\dfrac{\Gamma(\frac12)\Gamma(n+\frac12)}{\Gamma(n+1)}\pm B_{s^2}\left(\dfrac12, n+\dfrac12\right)\right]$

Source: Goodman [25].

[a] $s = -\gamma\cos(\phi-\phi_0)$. $B_{s^2}(\ ,\ )$ is the incomplete beta function. For the p.d.f. of $\hat\phi$ the plus sign applies if $|\hat\phi-\phi_0| \leq \pi/2$; the minus sign applies if $\pi/2 \leq |\hat\phi-\phi_0| \leq \pi$.

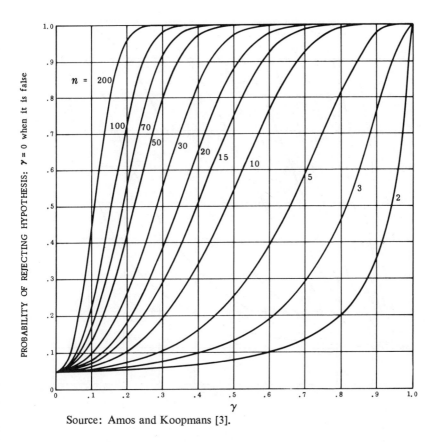

Source: Amos and Koopmans [3].

Fig. 24 Power curves for tests of hypothesis: $\gamma = 0$, 0.05 significance level

3.19. Testing the Zero Coherence Hypothesis

It is often of interest to test the hypothesis of zero coherence between two sequences. Table 9 lists the critical values for $\hat{\gamma}$ for the 5-, 10-, and 20-percent significance levels along with the expectation, variance, and mean-square error for several values of n. It is apparent that for small n, the mean-square error is dominated by the bias. To determine the power of the test for zero coherence, one may use the power curves in Figs. 24 and 25.

For zero coherence we also have

$$(3.32a) \qquad p_n = E(\hat{\gamma}) = (n-1)B(3/2, n-1) = \frac{(\pi)^{\frac{1}{2}}\Gamma(n)}{2\Gamma(n+\frac{1}{2})},$$

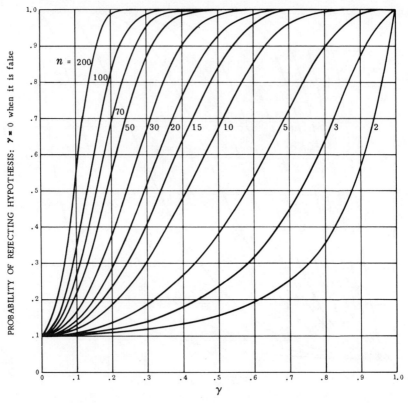

PROBABILITY OF REJECTING HYPOTHESIS: $\gamma = 0$ when it is false

Source: Amos and Koopmans [3].

Fig. 25 Power curves for tests of hypothesis: $\gamma = 0$, 0.10 significance level

$$(3.32b) \qquad q_n = \text{var}\,(\hat{\gamma}) = \frac{1}{n} - \pi\left[\frac{\Gamma(n)}{2\Gamma(n + \frac{1}{2})}\right]^2 \qquad n > 2.$$

While the hypothesis testing described above is interesting, it is often convenient to have a summary test statistic for the entire frequency interval. Suppose that we compute $(L + 1)$ estimates of γ on the interval $(0, \pi)$ such that

$$\text{cov}\,[\hat{\gamma}(\lambda_j),\, \hat{\gamma}(\lambda_k)] = 0,$$

$$\lambda_j = \pi j/L, \quad \lambda_k = \pi k/L, \quad j, k = 0, 1, \dots, L.$$

Under the zero coherence hypothesis it follows from the central limit theorem that†

---

† See also Nettheim [71, pp. 54–58].

$$(3.33) \qquad \Delta_L = \sum_{j=0}^{L} [\hat{\gamma}(\lambda_j) - p_{n_j}]/[(L+1)q_{n_j}]^{\frac{1}{2}}$$

$$= \left\{ \left[\sum_{j=1}^{L-1} \hat{\gamma}(\lambda_j) - (L-1)p_{2n_0} \right] \middle/ q_{2n_0}^{\frac{1}{2}} \right.$$

$$\left. + [\hat{\gamma}(0) + \hat{\gamma}(\pi) - 2p_{n_0}]/q_{n_0}^{\frac{1}{2}} \right\} \middle/ (L+1)^{\frac{1}{2}},$$

$$n_j = 1/(\delta_{\lambda_j}\Psi_{T,M}) = \varkappa_{\lambda_j}/\Psi_{T,M}$$

asymptotically has the normal distribution with zero mean and unit variance.

Table 9

Critical Values for Sample Coherence for Tests of the Hypothesis:

$\gamma = 0$, for α Significance Level

n	$\alpha = .05$	$\alpha = .10$	$\alpha = .20$	$E(\hat{\gamma})$	var $(\hat{\gamma})$	MSE $(\hat{\gamma})$
2	0.975	0.950	0.895	0.667	0.056	0.501
3	0.880	0.825	0.745	0.533	0.049	0.333
4	0.795	0.730	0.645	0.457	0.041	0.250
5	0.725	0.660	0.575	0.406	0.035	0.200
6	0.670	0.610	0.525	0.369	0.030	0.166
7	0.625	0.565	0.485	0.341	0.027	0.143
8	0.590	0.530	0.455	0.318	0.024	0.125
9	0.560	0.500	0.425	0.300	0.021	0.111
10	0.530	0.475	0.405	0.284	0.020	0.100

Source: Amos and Koopmans [3].

The spacing between estimates is generally such that they are positively correlated when the Parzen window is used. This certainly occurs when $L = M$; it also occurs to a lesser extent when $L = M/2$. Regarding Δ_L as having a unit variance then leads to an understatement of the confidence interval for a given significance level. Using Δ_L as a test statistic when estimates are correlated therefore increases the probability of rejecting the hypothesis, $\gamma = 0$, when it is true beyond the stated significance level. When the hypothesis is rejected it is wise to check the extent of correlation between estimates. If it is high then one may use more widely spaced estimates so as to reduce the upward bias in the rejection probability.

3.20. Approximate Confidence Intervals for Gain and Phase Angle

With regard to the sample frequency response function, it can be shown that for well-resolved estimates, the simultaneous confidence intervals for gain and phase angle are given by†

$$(3.34) \quad \text{Prob} \left[\begin{matrix} \hat{G}(1 + \sin \Delta\phi)^{-1} \leq G \leq \hat{G}(1 - \sin \Delta\phi)^{-1} \\ \hat{\phi} - \Delta\phi \leq \phi \leq \hat{\phi} + \Delta\phi \end{matrix} \right] \geq P,$$

$$\sin \Delta\phi = \{[1 - \gamma^2][(1 - P)^{-2/n} - 1]\}^{\frac{1}{2}}/\gamma.$$

These results are approximate, since they are a consequence of the results in Section 3.16, and also depend for their validity on the input sequences (1) being measured without error and (2) being uncorrelated with the residual sequence that makes up Y.

3.21. Multivariate Estimation Procedures

Estimation for multivariate time series follows bivariate procedures to a great extent. Using the notation of Section 2.30, we consider a sample of T vector observations on the sets of sequences $\{X_t\}$ and $\{Y_t\}$, where X_t is an $N \times 1$ column vector of input sequences and Y_t is a $p \times 1$ column vector of output sequences. The sample vector means are

$$(3.35a) \qquad \bar{X} = T^{-1} \sum_{t=1}^{T} X_t, \qquad \bar{Y} = T^{-1} \sum_{t=1}^{T} Y_t,$$

and the sample covariance matrices are

$$(3.35b) \quad \begin{cases} C_{xx,\tau} = T^{-1} \sum_{t=1}^{T-|\tau|} [(X_t - \bar{X})(X_{t+\tau} - \bar{X})'] \\[2mm] C_{yy,\tau} = T^{-1} \sum_{t=1}^{T-|\tau|} [(Y_t - \bar{Y})(Y_{t+\tau} - \bar{Y})'] \\[2mm] C_{xy,\tau} = T^{-1} \sum_{t=1}^{T-|\tau|} [(X_t - \bar{X})(Y_{t+\tau} - \bar{Y})']. \end{cases}$$

Hereafter we concentrate on the sampling properties of $C_{xy,\tau}$ since those for $C_{xx,\tau}$ and $C_{yy,\tau}$ are analogous.

Assuming T sufficiently large, the bias is

$$(3.36) \qquad E(C_{xy,\tau} - R_{xy,\tau}) \sim -2\pi g_{xy}(0)/T.$$

To derive the asymptotic covariance matrix, a rearrangement of columns is necessary. Let $P_{xy,\tau}$ be an $Np \times 1$ vector with $R_{x_j y_k, \tau}$ in

---

† See Goodman [25, pp. 136–137 and 159–160].

row $p(j - 1) + k$. As an estimator of $\mathbf{P}_{xy,\tau}$, we take

$$(3.37) \qquad \mathbf{D}_{xy,\tau} = T^{-1} \sum_{t=1}^{T-|\tau|} [(\mathbf{X}_t - \overline{\mathbf{X}}) \otimes (\mathbf{Y}_{t+\tau} - \overline{\mathbf{Y}})],$$

where $(\mathbf{A} \otimes \mathbf{B})$ is the Kronecker product of $\mathbf{A}$ and $\mathbf{B}$† and $C_{x_j y_k, \tau}$ is in row $p(j - 1) + k$ of the $Np \times 1$ vector $\mathbf{D}_{xy,\tau}$.

Assuming that $\{\mathbf{X}_t\}$ and $\{\mathbf{Y}_t\}$ have a joint normal distribution, then

$$(3.38a) \qquad E(X_{j,t} X_{k,\tau} Y_{l,s} Y_{m,\nu}) = R_{x_j x_k, \tau - t} R_{y_l y_m, \nu - s}$$

$$+ R_{x_j y_l, s-t} R_{x_j y_m, \nu - \tau}$$

$$+ R_{x_j y_m, \nu - \tau} R_{x_k y_l, s - \tau}.$$

Using the property

$$(3.38b) \qquad (\mathbf{X}_t \otimes \mathbf{Y}_{t+\tau})(\mathbf{X}_s \otimes \mathbf{Y}_{s+\nu})' = (\mathbf{X}_t \mathbf{X}_s') \otimes (\mathbf{Y}_{t+\tau} \mathbf{Y}_{s+\nu}'),$$

we have

$$(3.38c) \qquad E\{[(\mathbf{X}_t \otimes \mathbf{Y}_{t+\tau}) - \mathbf{P}_{xy,\tau}][(\mathbf{X}_s \otimes \mathbf{Y}_{s+\nu}) - \mathbf{P}_{xy,\nu}]'\}$$

$$= (\mathbf{R}_{xx,s-t} \otimes \mathbf{R}_{yy,s-t+\nu-\tau}) + \mathbf{Q}_{xy,\tau+t-s,\nu+s-t},$$

where the $Np \times Np$ matrix $\mathbf{Q}_{xy,\tau+t-s,\nu+s-t}$ has the $Np \times p$ matrix $(\mathbf{R}_{xy,\tau+t-s} \otimes \mathbf{R}'_{x_j y, \nu+s-t})$ in columns $p(j - 1) + 1$ through pj, $\mathbf{R}'_{x_j y, \nu+s-t}$ being the transpose of the jth row of $\mathbf{R}_{xy, \nu+s-t}$.

Noting that the Fourier transform of $\mathbf{R}_{x_j y, \tau}$ is $\mathbf{g}_{x_j y}(\omega)$, we see that the asymptotic covariance matrix of (3.37) is

$$(3.38d) \qquad E[(\mathbf{D}_{xy,\tau} - \mathbf{P}_{xy,\tau})(\mathbf{D}_{xy,\nu} - \mathbf{P}_{xy,\nu})']$$

$$\sim T^{-1} \sum_{v=-\infty}^{\infty} [(\mathbf{R}_{xx,v} \otimes \mathbf{R}_{yy,v+\nu-\tau}) + \mathbf{Q}_{xy,\tau-v,\nu+v}]$$

$$= 2\pi T^{-1} \int_{-\pi}^{\pi} \{[\mathbf{g}_{xx}(\omega) \otimes \mathbf{g}_{yy}(\omega)]e^{i\omega(\nu-\tau)}$$

$$+ \mathbf{H}_{xy}(\omega)e^{i\omega(\tau+\nu)}\} \, d\omega,$$

where the $Np \times Np$ matrix $\mathbf{H}_{xy}(\omega)$ has the $Np \times p$ matrix $[\mathbf{g}_{xy}(\omega) \otimes \mathbf{g}'_{x_j y}(\omega)]$ in columns $p(j - 1) + 1$ through pj.

The estimator of $\mathbf{g}_{xy}(\lambda)$ is

$$(3.39a) \qquad \hat{\mathbf{g}}_{xy}(\lambda) = (2\pi)^{-1} \sum_{\tau=-M}^{M} \mathbf{C}_{xy,\tau} k_{M,\tau} e^{-i\lambda\tau}$$

---

† See Halmos [34, pp. 95–96].

which, in column notation, is

$$(3.39b) \qquad \hat{\mathbf{h}}_{xy}(\lambda) = (2\pi)^{-1} \sum_{\tau=-M}^{M} \mathbf{D}_{xy,\tau} k_{M,\tau} e^{-i\lambda\tau}.$$

Here $\hat{\mathbf{h}}_{xy}(\lambda)$ is an $Np \times 1$ vector with $\hat{g}_{x_j y_k}(\lambda)$ in row $p(j-1)+k$. We write

$$\mathbf{h}_{xy}(\lambda) = \mathbf{h}_{xy}^c(\lambda) - i\mathbf{h}_{xy}^q(\lambda),$$

where $\mathbf{h}_{xy}^c(\lambda)$ and $\mathbf{h}_{xy}^q(\lambda)$ are the column vectors of co-spectra and quadrature-spectra respectively. The asymptotic bias in (3.39b) is analogous to that for the bivariate case. It is

$$(3.40a) \quad E[\hat{\mathbf{h}}_{xy}^c(\lambda)] \sim \int_{-\pi}^{\pi} K_M(\lambda - \omega)\mathbf{h}_{xy}^c(\omega)\,d\omega - 2\pi\mathbf{h}_{xy}^c(0)K_M(\lambda)/T,$$

$$(3.40b) \quad E[\hat{\mathbf{h}}_{xy}^q(\lambda)] \sim \int_0^\pi [K_M(\lambda - \omega) - K_M(\lambda + \omega)]\mathbf{h}_{xy}^q(\omega)\,d\omega.$$

The bias corrections described in Section 3.14 also apply here.

To derive the asymptotic covariance properties, we note that

$$(3.40c) \quad E\{[\hat{\mathbf{h}}_{xy}(\lambda) - \mathbf{h}_{xy}(\lambda)][\hat{\mathbf{h}}_{xy}(\lambda') - \mathbf{h}_{xy}(\lambda')]'\}$$

$$\sim 2\pi T^{-1} \int_{-\pi}^{\pi} \{[\mathbf{g}_{xx}(\omega) \otimes \mathbf{g}_{yy}(\omega)]K_M(\lambda + \omega)$$

$$+ \mathbf{H}_{xy}(\omega)K_M(\lambda - \omega)\} \times K_M(\lambda' - \omega)\,d\omega,$$

$$(3.40d) \quad E\{[\hat{\mathbf{h}}_{xy}(-\lambda) - \mathbf{h}_{xy}(-\lambda)][\hat{\mathbf{h}}_{xy}(\lambda') - \mathbf{h}_{xy}(\lambda')]'\}$$

$$\sim 2\pi T^{-1} \int_{-\pi}^{\pi} \{[\mathbf{g}_{xx}(\omega) \otimes \mathbf{g}_{yy}(\omega)]K_M(\lambda - \omega)$$

$$+ \mathbf{H}_{xy}(\omega)K_M(\lambda + \omega)\} \times K_M(\lambda' - \omega)\,d\omega.$$

Then for well-resolved estimates, we have for $0 < |\lambda| < \pi$

$$(3.40e) \quad \left\{ \begin{array}{l} \text{cov}\,[\hat{\mathbf{h}}_{xy}^c(\lambda),\, \hat{h}_{xy}^{c'}(\lambda)] \\ \text{cov}\,[\hat{\mathbf{h}}_{xy}^q(\lambda),\, \hat{\mathbf{h}}_{xy}^{q'}(\lambda)] \\ \text{cov}\,[\hat{h}_{xy}^c(\lambda),\, \hat{\mathbf{h}}_{xy}^{q'}(\lambda)] \\ \text{cov}\,[\hat{\mathbf{h}}_{xy}(-\lambda),\, \hat{\mathbf{h}}_{xy}'(\lambda)] \end{array} \right\}$$

$$\sim \frac{1}{2}\, \Psi_{T,M} \left\{ \begin{array}{c} [\mathbf{c}_{xx}(\lambda) \otimes \mathbf{c}_{yy}(\lambda)] - [\mathbf{q}_{xx}(\lambda) \otimes \mathbf{q}_{yy}(\lambda)] + \mathbf{H}_{xy}^c(\lambda) \\ [\mathbf{c}_{xx}(\lambda) \otimes \mathbf{c}_{yy}(\lambda)] - [\mathbf{q}_{xx}(\lambda) \otimes \mathbf{q}_{yy}(\lambda)] - \mathbf{H}_{xy}^c(\lambda) \\ [\mathbf{c}_{xx}(\lambda) \otimes \mathbf{q}_{yy}(\lambda)] + [\mathbf{q}_{xx}(\lambda) \otimes \mathbf{c}_{yy}(\lambda)] + \mathbf{H}_{xy}^q(\lambda) \\ 2[\mathbf{g}_{xx}(\lambda) \otimes \mathbf{g}_{yy}(\lambda)] \end{array} \right\},$$

where

$$\mathbf{g}_{xx}(\lambda) = \mathbf{c}_{xx}(\lambda) - i\mathbf{q}_{xx}(\lambda)$$

$$\mathbf{g}_{yy}(\lambda) = \mathbf{c}_{yy}(\lambda) - i\mathbf{q}_{yy}(\lambda)$$

$$\mathbf{g}_{xy}(\lambda) = \mathbf{c}_{xy}(\lambda) - i\mathbf{q}_{xy}(\lambda)$$

$$\mathbf{H}_{xy}(\lambda) = \mathbf{H}_{xy}^c(\lambda) - i\mathbf{H}_{xy}^q(\lambda).$$

Here the $Np \times Np$ matrices $\mathbf{H}_{xy}^c(\lambda)$ and $\mathbf{H}_{xy}^q(\lambda)$ have $\{[\mathbf{c}_{xy}(\lambda) \otimes \mathbf{c}'_{x_jy}(\lambda)] - [\mathbf{q}_{xy}(\lambda) \otimes \mathbf{q}'_{x_jy}(\lambda)]\}$ and $\{[\mathbf{q}_{xy}(\lambda) \otimes \mathbf{c}'_{x_jy}(\lambda)] + [\mathbf{c}_{xy}(\lambda) \otimes \mathbf{q}'_{x_jy}(\lambda)]\}$, respectively, in columns $p(j-1) + 1$ through pj. The estimation of $\hat{\mathbf{g}}_{xx}(\lambda)$ and $\hat{\mathbf{g}}_{yy}(\lambda)$, their asymptotic biases, and their covariance matrices follow from (3.37) through (3.40) by the appropriate substitution of variables and subscripts. For $|\lambda| = 0, \pi$ the covariances are double those given in (3.30) except for cov $[\hat{\mathbf{h}}_{xy}^q(\lambda), \hat{\mathbf{h}}_{xy}^{q'}(\lambda)]$ and cov $[\hat{\mathbf{h}}_{xy}^c(\lambda), \hat{\mathbf{h}}_{xy}^{q'}(\lambda)]$ which are zero.

For the set of simultaneous equations described by (2.49), we have

(3.41) $$\hat{\mathbf{g}}_{\eta\eta}(\lambda) = \hat{\mathbf{g}}_{yy}(\lambda) - \hat{\mathbf{g}}_{xy}^*(\lambda)\hat{\mathbf{g}}_{xx}^{-1}(\lambda)\hat{\mathbf{g}}_{xy}(\lambda).$$

This $p \times p$ spectrum matrix estimator plays a major role in estimating the coefficients in (2.49), as described in Chap. 4. In the next section we shall describe multivariate tests that can be performed on $\hat{\mathbf{g}}_{\eta\eta}(\lambda)$ in order to simplify further analysis.

For the single distributed lag equation (2.43), we obtain the estimators of multiple and partial coherences, and the joint frequency response:

(3.42a) $$\hat{\gamma}_{y \cdot 1, 2, \ldots, N}(\lambda) = \{[\hat{\mathbf{g}}_{xy}^*(\lambda)\hat{\mathbf{g}}_{xx}^{-1}(\lambda)\hat{\mathbf{g}}_{xy}(\lambda)]/\hat{g}_y(\lambda)\}^{\frac{1}{2}},$$

(3.42b) $$\hat{\gamma}_{y \cdot p+1, p+2, \ldots, N | 1, 2, \ldots, p}(\lambda) = \{[\hat{\mathbf{g}}_{\xi y}^*(\lambda)\hat{\mathbf{g}}_{\xi\xi}^{-1}(\lambda)\hat{\mathbf{g}}_{\xi y}(\lambda)]/\hat{g}_y(\lambda)\}^{\frac{1}{2}},$$

(3.42c) $$\hat{\mathbf{g}}_{\xi\xi}(\lambda) = \hat{\mathbf{h}}_{22}(\lambda) - \hat{\mathbf{h}}_{21}(\lambda)\hat{\mathbf{h}}_{11}^{-1}(\lambda)\hat{\mathbf{h}}_{12}(\lambda),$$

(3.42d) $$\hat{\mathbf{g}}_{\xi y}(\lambda) = \hat{\mathbf{m}}_2(\lambda) - \hat{\mathbf{h}}_{21}(\lambda)\hat{\mathbf{h}}_{11}^{-1}(\lambda)\hat{\mathbf{m}}_1(\lambda),$$

(3.42e) $$\hat{\mathbf{A}}(\lambda) = \hat{\mathbf{g}}_{xx}^{-1}(\lambda)\hat{\mathbf{g}}_{xy}(\lambda)$$

respectively. The circumflexes over the matrices indicate the substitution of estimators for the population measures defined in Section 2.30. The approximating distributions for multiple and partial coherence are given in [26] and [28], as are the simultaneous confidence intervals for the matrix frequency response function. Tests related to the estimators in Expressions (3.42a–e) are described in [7].

A word of caution is called for regarding the estimation and manipulation of these multivariate matrices. The matrix operations described here are performed on matrices with complex elements and therefore must be performed with special care. This, along with the necessity of performing these operations at a number of frequencies, makes the required matrix inversions time consuming even on a high-speed digital computer. The investigator intending to use these results is well advised to use proven computer programs especially designed to accomplish these operations.

For convenience of exposition, we assumed that all sample record lengths T are of the same length and the number of lags M used in spectrum estimation is the same for all spectra. The quantity

$$n_\lambda = (\delta_\lambda \Psi_{T,M})^{-1}$$

is therefore the same for all sample cross spectra.

3.22. The Joint Distribution of Multivariate Spectrum Estimates

As in the bivariate case, the approximating distribution theory for multivariate spectrum analysis relies heavily on Goodman [26, 27]. For well-resolved estimates, the elements of the sample spectral matrix

$$n\hat{\mathbf{g}}(\lambda) = n \begin{bmatrix} \hat{\mathbf{g}}_{xx}(\lambda) & \hat{\mathbf{c}}_{xy}(\lambda) - i\hat{\mathbf{q}}_{xy}(\lambda) \\ \hat{\mathbf{c}}'_{xy}(\lambda) + i\hat{\mathbf{q}}'_{xy}(\lambda) & \hat{\mathbf{g}}_{yy}(\lambda) \end{bmatrix},$$

$$n = 1/\Psi_{T,M} \geq N + p$$

jointly have the complex Wishart distribution with p.d.f.

$$w_C[n\hat{\mathbf{g}}(\lambda)|\mathbf{g}(\lambda), n] \quad \text{for} \quad 0 < |\lambda|\pi$$

and have the Wishart distribution with p.d.f.

$$w_R[n\hat{\mathbf{g}}(\lambda)|\mathbf{g}(\lambda), n] \quad \text{for} \quad |\lambda| = 0, \pi.$$

The statistic $n_\lambda^{N+p}|\hat{\mathbf{g}}(\lambda)|/|\mathbf{g}(\lambda)|$ has the distribution of the product of $N + p$ independent chi-square variates with n_λ, $(n_\lambda - n_\lambda)$, $\ldots$, $[n_\lambda - (N + p - 1)n_\lambda]$ equivalent degrees of freedom respectively, for $0 \leq |\lambda| \leq \pi$. We may also write

$$|\hat{\mathbf{g}}(\lambda)| = |\hat{\mathbf{g}}_{yy}(\lambda)\hat{\mathbf{g}}_{xx}(\lambda) - \hat{\mathbf{g}}^*_{xy}(\lambda)\hat{\mathbf{g}}_{xy}(\lambda)|$$

$$= |\hat{\mathbf{g}}_{yy}(\lambda) - \hat{\mathbf{g}}^*_{xy}(\lambda)\hat{\mathbf{g}}_{xx}^{-1}(\lambda)\hat{\mathbf{g}}_{xy}(\lambda)| \cdot |\hat{\mathbf{g}}_{xx}(\lambda)|$$

$$= |\hat{\mathbf{g}}_{\eta\eta}(\lambda)| \cdot |\hat{\mathbf{g}}_{xx}(\lambda)|.$$

Then we treat the statistic $n_\lambda^N |\hat{\mathbf{g}}_{xx}(\lambda)| / |\mathbf{g}_{xx}(\lambda)|$ as having the distribution of the product of N independent chi-square variates with n_λ, $(n_\lambda - \varkappa_\lambda)$, $\ldots$, $[n_\lambda - (N - 1)\varkappa_\lambda]$ equivalent degrees of freedom, respectively, and the statistic $n_\lambda^p |\hat{\mathbf{g}}_{\eta\eta}(\lambda)| / |\mathbf{g}_{\eta\eta}(\lambda)|$ as having the distribution of the product of p independent chi-square variates with $(n_\lambda - N\varkappa_\lambda) - [n_\lambda - (N + 1)\varkappa_\lambda], \ldots, [n_\lambda - (N + p - 1)\varkappa_\lambda]$ equivalent degrees of freedom respectively.

As in the bivariate case the joint distribution of spectrum and cross spectrum converges to the multivariate normal distribution, a result that proves helpful in developing a distribution theory for estimated coefficients in simultaneous equation models in Chap. 4.

The approximate distribution theory for $\hat{\mathbf{g}}_{\eta\eta}(\lambda)$ is particularly convenient for it enables us to test hypotheses about the residual or conditional spectral matrix $\mathbf{g}_{\eta\eta}(\lambda)$ within the framework of the multivariate testing procedures described in Anderson [4]. The next section discusses several likelihood ratio tests that should assist investigators during the model-building phase of an econometric study.

3.23. Multivariate Testing Procedures

In Section 2.30 we stressed the value of the information that the spectrum matrix $\mathbf{g}_{\eta\eta}(\lambda)$ can provide. For example, if $\mathbf{g}_{\eta\eta}(\lambda)$ is a diagonal matrix for all λ in $(-\pi, \pi)$, then we may estimate the coefficients of each equation defined by (2.49) separately from those of the remaining equations without any loss of efficiency. If we can additionally establish that $\mathbf{g}_{\eta\eta}(\lambda)$ is constant over $(-\pi, \pi)$, then the unweighted linear least-squares method applied to the individual equations provides best linear unbiased estimates.

In the same manner as described in Section 3.22, we approximate the distribution of $(n - N)\hat{\mathbf{g}}_{\eta\eta}(\lambda)$ by the complex Wishart distribution for $0 < |\lambda| < \pi$ and by the Wishart distribution when $|\lambda| = 0, \pi$. In Anderson [4] a number of likelihood ratio tests are given for real normal random variables with independent observations. If we treat the case of $0 < |\lambda| < \pi$ as equivalent to collecting $2(n - N)$ independent observations and the case of $|\lambda| = 0, \pi$ as equivalent to collecting $n - N$ independent observations, it seems reasonable to use the aforementioned likelihood ratio tests, properly adjusted, as rough guides in empirical investigation.

Suppose we wish to test the hypothesis that the residual sequences $\{\boldsymbol{\eta}_t\}$ are mutually independent. This amounts to testing the hypothesis that $\mathbf{g}_{\eta\eta}(\lambda)$ is a diagonal matrix for all λ in $(-\pi, \pi)$. Under the null hypothesis, the likelihood ratio for frequency λ leads to the test statistic

$$(3.43a) \qquad V_1(\lambda) = |\hat{\mathbf{g}}_{\eta\eta}(\lambda)| \Big/ \prod_{j=1}^{p} [\hat{\mathbf{g}}_{\eta_j\eta_j}(\lambda)]$$

whose distribution has an asymptotic expansion that yields†

$$(3.43b) \qquad \Pr[-m_\lambda \log V_1(\lambda) \le v] \sim \Pr(\chi_f^2 \le v) + \frac{\gamma_2}{m_\lambda^2}[\Pr(\chi_{f+4}^2 \le v)$$

$$- \Pr(\chi_f^2 \le v)],$$

$$(3.43c) \qquad\qquad f = p(p-1)/2,$$

$$(3.43d) \qquad\qquad m_\lambda = (n-N)\varkappa_\lambda - (2p+11)/6,$$

$$(3.43e) \qquad\qquad \gamma_2 = p(p-1)(2p^2 - 2p - 13)/288.$$

Since

$$\mathbf{g}_{\eta\eta}(\lambda) = \mathbf{g}_{\eta\eta}^*(-\lambda),$$

we need only concern ourselves with the interval $(0, \pi)$.

If we have $L + 1$ independent estimates in $(0, \pi)$, then we may approximate the distribution of

$$-\sum_{j=0}^{L} m_{\lambda_j} \log V_1(\lambda_j), \qquad \lambda_j = \pi j/L$$

by that of chi-square with $(L + 1)f$ degrees of freedom.

Suppose we wish to test the hypothesis that $\{\eta_t\}$ is a set of p white noise processes so that $\mathbf{g}_{\eta\eta}(\lambda)$ is constant over $(-\pi, \pi)$. In the language of multivariate analysis, this amounts to testing the hypothesis

$$\mathbf{g}_{\eta\eta}(\lambda_0) = \mathbf{g}_{\eta\eta}(\lambda_1) = \cdots = \mathbf{g}_{\eta\eta}(\lambda_L)$$

for $(L + 1)$ independent estimates on the frequency interval $(0, \pi)$. The test statistic is‡

$$(3.44a) \qquad\qquad W_2 = V_2 \prod_{j=0}^{L} (n/n_j)^{pn_j/2},$$

where

$$(3.44b) \qquad V_2 = \left[\prod_{j=0}^{L} |(n_j+1)\hat{\mathbf{g}}_{\eta\eta}(\lambda_j)|^{n_j/2} \right] \Big/ \left| 4n_0 L \sum_{j=0}^{L} \hat{\mathbf{g}}_{\eta\eta}(\lambda_j) \right|^n,$$

$$(3.44c) \qquad n_j = (n-N)\varkappa_{\lambda_j} - 1,$$

---

† See Anderson [4, p. 239].
‡ See Anderson [4, Chap. 10].

(3.44d) $\qquad n = \sum_{j=0}^{L} n_j = 4n_0 L - (L + 1),$

with approximating distribution such that

(3.44e) $\qquad \Pr(-2\rho \log W_2 \leq w)$
$$\sim \Pr(\chi_f^2 \leq w) + \omega_2[\Pr(\chi_{f+4}^2 \leq w) - \Pr(\chi_f^2 \leq w)],$$

(3.44f) $\qquad \rho = 1 - \left[\sum_{j=0}^{L} (1/n_j) - 1/n\right][2p^2 + 3p - 1]/[6L(p + 1)],$

(3.44g) $\qquad \omega_2 = p(p + 1)\left\{(p - 1)(p + 2)\left[\sum_{j=0}^{L} (1/n_j)^2 - 1/n^2\right]\right.$
$$\left. - 6L(1 - \rho)^2\right\}\Big/ (48\rho^2),$$

(3.44h) $\qquad f = Lp(p + 1)/2.$

We now consider testing the hypothesis that $\{\eta_t\}$ is a set of mutually independent white noise processes. We proceed as in the preceding test with the added assumption that $\mathbf{g}_{\eta\eta}(\lambda)$ is a diagonal matrix. For $L + 1$ independent estimates, the test statistic is

(3.45a) $\qquad W_3 = V_3 \prod_{j=0}^{L} (n/n_j)^{pn_j/2},$

(3.45b) $V_3 = \prod_{j=0}^{L} |(n_j + 1)\hat{\mathbf{g}}_{\eta\eta}(\lambda_j)|^{n_j/2} \Big/ \left\{4n_0 L \prod_{k=1}^{p}\left[\sum_{j=0}^{L} g_{\eta_k \eta_k}(\lambda_j)\right]\right\}^{n/2},$

with approximating distribution such that

$$\Pr(-2\rho \log W_3 \leq w) \sim \Pr(\chi_f^2 \leq w)$$
$$+ \omega_2[\Pr(\chi_{f+4}^2 \leq w) - \Pr(\chi_f^2 \leq w)].$$

The quantities n_j, n, ρ, ω_2 and f are defined by (3.44c), (3.44d), (3.44f), (3.44g), and (3.44h), respectively.

The above testing procedures rely on a number of approximations and heuristic arguments and should therefore be considered as rough guides rather than as literal truths. The value of the procedures lies in their ability to reduce a multidimensional problem to one of testing a single number. In an exploratory investigation, these procedures, although rough, permit informed decision making about how sophisticated the ensuing analysis should be. In Section 4.13, we shall describe multivariate testing procedures as they apply to the analysis of residuals.

4 Distributed Lag Models

This chapter describes spectral methods for estimating and testing distributed lag models. The procedures are primarily due to Hannan [38,39,43] and have been used in economics by Wallis [96,97]† to study inventory problems.

While frequency-domain statistics provide useful insights into behavior and association among time-varying phenomena, a principal purpose of many econometric studies is to construct a time-domain model that explains one set of economic phenomena by another. The distributed lag model described in Section 2.30 is particularly appealing for this purpose, for it enables the investigator to express time lags explicitly. Since the response of one economic phenomena to a change in another is seldom instantaneous, the inclusion of lagged variables becomes highly desirable if a comprehensive picture of economic interactions is to be obtained.

In most econometric studies using distributed lag models, lagged variables are regarded as individual, though correlated, entities. This approach is a consequence of studying these models within the general estimation and testing framework of multivariate analysis. In contrast, the spectral approach regards lagged variables as the elements of a

---

† Hamon and Hannan [35] present an interesting and instructive application of these procedures to a problem in oceanography. See Rosenblatt [79] for earlier work on regression of time series and Parzen [77, 78] and Wahba [94] for a related approach to estimating coefficients in a distributed lag model.

covariance stationary sequence, thereby exploiting properties of these sequences that simplify estimation and testing. Foremost among the simplifying properties is the asymptotic behavior of the covariance matrix of a covariance stationary time series when the matrix is subjected to a unitary transformation.

In his comprehensive book on econometrics [62], Malinvaud describes approaches to estimating coefficients in a distributed lag model and cites three problems that impair successful estimation. They are multicollinearity among the lagged variables, autocorrelated residuals, and the choice of the appropriate number of lags. We raise these problems again to show how they may be understood in spectral terms.

We shall describe the bivariate case first, and then present the more complicated multivariate generalization.

4.1. Problem Definition

Consider the distributed lag model

$$(4.1) \qquad Y_t = \sum_{s=-r_1}^{r_2} a_s X_{t-s} + \eta_t,$$

where X, Y, and η are stationary normal sequences with zero means. Given T pairs of observations on X and Y, the problem is to estimate the vector

$$\mathbf{a} = [a_{-r_1}, a_{-r_1+1}, \ldots, a_0, \ldots, a_{r_2}].$$

Eight points of clarification are worthwhile making before developing the procedure.

1. We assume X and η are independent so that

$$g_{x\eta}(\lambda) = 0.$$

If in the estimation model r_1 and r_2 were incorrectly specified, then a part of the association between X and Y would reside in the residual process, thereby creating a dependence between that process and X.

2. In econometrics, it is customary to regard the endogenous variable Y_t as a response to $X_t, X_{t-1}, \ldots, X_{t-s}$. For the model to hold statistically, however, it may be necessary to include negative lags. In Section 4.11 we shall discuss how the presence of negative lags can be interpreted in a way consistent with economic theory.

3. The residual spectrum g_η is positive in $(-\pi, \pi)$ so that covariance matrix of η is positive definite and hence nonsingular.

4. From Expression (4.1) we note that

$$C_{\eta,\tau} = C_{y,\tau} + \sum_{j,s=-r_1}^{r_2} a_j a_s C_{x,\tau-j+s} - \sum_{s=-r_1}^{r_2} a_s(C_{xy,\tau+s} + C_{xy,-\tau+s}),$$

ignoring end effects due to finite T. We then have the periodogram,

(4.2a)

$$I_\eta(\lambda) = I_y(\lambda) + \sum_{j,s=-r_1}^{r_2} a_j a_s I_{x_{-j}x_s}(\lambda) - 2 \sum_{s=-r_1}^{r_2} a_s \operatorname{Re}[I_{x_0 y_s}(\lambda)],$$

and the spectrum estimator,

(4.2b)

$$\hat{g}_\eta(\lambda) = \hat{g}_y(\lambda) + \sum_{j,s=-r_1}^{r_2} a_j a_s \hat{g}_{x_{-j}x_s}(\lambda) - 2 \sum_{s=-r_1}^{r_2} a_s \operatorname{Re}[\hat{g}_{x_0 y_s}(\lambda)],$$

where Re [ ] denotes the real part of the quantity in brackets and

(4.2c)
$$I_{x_{-j}x_s}(\lambda) = (1/T) \sum_{\tau=-T+1}^{T-1} C_{x,\tau-j+s} e^{-i\lambda\tau},$$

(4.2d)
$$I_{x_0 y_s}(\lambda) = (1/T) \sum_{\tau=-T+1}^{T-1} C_{xy,\tau+s} e^{-i\lambda\tau},$$

(4.2e)
$$\hat{g}_{x_{-j}x_s}(\lambda) = (2\pi)^{-1} \sum_{\tau=-M}^{M} k_{M,\tau} C_{x,\tau-j+s} e^{-i\lambda\tau},$$

(4.2f)
$$\hat{g}_{x_0 y_s}(\lambda) = (2\pi)^{-1} \sum_{\tau=-M}^{M} k_{M,\tau} C_{xy,\tau+s} e^{-i\lambda\tau}.$$

These expressions are fundamental to the estimation of **a**. In their present forms they are cumbersome, owing to the necessity of computing them for all j and s. A problem of resolution also arises as we shall demonstrate using Expression (4.2e). Centering the weighting function k on a sample autocovariance other than $C_{x,0}$ impairs our ability to obtain well-resolved estimates. Notice that

$$E[\hat{g}_{x_{-j}x_s}(\lambda)] \sim (2\pi)^{-1} \sum_{\tau=-M}^{M} k_{M,\tau} R_{x,\tau-j+s} e^{-i\lambda\tau}$$

$$= \int_{-\pi}^{\pi} K_M(\lambda - \omega) g_x(\omega) e^{i\omega(s-j)} d\omega$$

so that $g_x(\omega)e^{i\omega(s-j)}$ oscillates as ω varies in $(-\pi, \pi)$, the period of oscillation being the reciprocal of $(s-j)$. To resolve this function properly, M must be larger than is necessary to resolve the spectrum g_x. This difficulty is analogous to the problem that arose in estimating the cross spectrum in Section 3.14.

As before, it is convenient to replace $\hat{g}_{x_{-j}x_s}(\lambda)$ by $\hat{g}_x(\lambda)e^{i\lambda(s-j)}$, the error of approximation being

$$\Delta(\lambda) = \hat{g}_{x_{-j}x_s}(\lambda) - \hat{g}_x(\lambda)e^{i\lambda(s-j)}$$

$$= (2\pi)^{-1} \sum_{\tau=-M}^{M} k_{M,\tau}[C_{x,\tau-j+s} - C_{x,\tau}e^{i\lambda(s-j)}]e^{-i\lambda\tau}$$

so that

$$E[\Delta(\lambda)] \sim \int_{-\pi}^{\pi} K_M(\lambda - \omega)g_x(\omega)[e^{i\omega(s-j)} - e^{i\lambda(s-j)}]\,d\omega.$$

This error clearly converges to zero as the averaging kernel K becomes more concentrated around the point $\omega = \lambda$. We also replace $\hat{g}_{x_0 y_s}(\lambda)$ by $\hat{g}_{xy}(\lambda)e^{i\lambda s}$ using similar arguments to those above. Here we also assume that $R_{xy,\tau}$ has a maximum at $\tau = 0$ so that we need not shift the weighting kernel as described in Section 3.14.

Since these results also hold for the periodogram, we replace Expressions (4.2a) and (4.2b) by

(4.3a) $$I_\eta(\lambda) = I_y(\lambda) + \sum_{j,s=-r_1}^{r_2} a_j a_s I_x(\lambda)e^{i\lambda(s-j)}$$

$$- 2 \sum_{s=-r_1}^{r_2} a_s \operatorname{Re}[I_{xy}(\lambda)e^{i\lambda s}],$$

(4.3b) $$\hat{g}_\eta(\lambda) = \hat{g}_y(\lambda) + \sum_{j,s=-r_1}^{r_2} a_j a_s \hat{g}_x(\lambda)e^{i\lambda(s-j)}$$

$$- 2 \sum_{s=-r_1}^{r_2} a_s \operatorname{Re}[\hat{g}_{xy}(\lambda)e^{i\lambda s}].$$

5. The approximation

(4.4) $$\hat{g}_{x\eta}(\lambda) = \hat{g}_{xy}(\lambda) + \hat{g}_x(\lambda) \sum_{s=-r_1}^{r_2} a_s e^{-i\lambda s}$$

is also taken to be valid by the same argument used in Point 4.

6. We assume sampling fluctuations are small enough to permit the use of the truncated Taylor series expansion,

$$(4.5) \qquad y/x \sim b/a + (y - b)/a - b(x - a)/a^2,$$

when deriving the variance of ratio estimators.

7. We recall that for sufficiently large T/M, we may regard cross-spectrum estimates as being approximately normally distributed. The use of this result requires at least 30 equivalent degrees of freedom, a somewhat unrealistic number for economic time series in general. When testing the validity of the model, however, one may resort to the tests described in Sections 3.19 and 3.23 and to additional criteria described in Section 4.10. These tests do not rely so heavily on the asymptotic result.

8. As mentioned earlier, it is often convenient to prewhiten X and Y prior to estimating their spectra. As was shown in Section 2.28, the estimated distributed lag coefficients apply equally to the original time series X and Y and to their corresponding prewhitened time series, provided that the same prewhitening procedure is applied to both X and Y.

4.2. The Estimation of Coefficients

We begin by considering the form of the maximum likelihood estimator. In an analysis based on independent observations, the likelihood function is the product of the density functions corresponding to the individual observations or to transformations of the individual observations. As mentioned earlier, in time series analysis we have one sample record on the stochastic process X and one sample record on the stochastic process Y. Using Expression (4.1), we may transform the joint distribution of the elements in the sample records on X and Y into the joint distribution of the residual elements $\eta_1, \eta_2, \ldots, \eta_T$, ignoring end effects.

Let η_T be the $T \times 1$ vector such that

$$\eta_T' = (\eta_1, \eta_2, \ldots, \eta_T).$$

Then the probability density function of η_T is

$$f(\eta_T) = \frac{\exp\left(-\tfrac{1}{2}\eta_T'\Gamma^{-1}\eta_T\right)}{(2\pi)^{T/2}|\Gamma|^{\frac{1}{2}}}$$

where Γ is the positive definite covariance matrix

$$\Gamma = \begin{bmatrix} R_{\eta,0} & R_{\eta,1} & \cdots & R_{\eta,T-1} \\ R_{\eta,-1} & R_{\eta,0} & & \\ \vdots & & \ddots & \\ R_{\eta,1-T} & \cdots & & R_{\eta,0} \end{bmatrix}.$$

Since we have but one sample record jointly on X and Y, f is also the likelihood function for the T pairs of dependent observations. Our purpose is to find the vector $\mathbf{a}$ that maximizes f or, more literally that maximizes the logarithm of the likelihood function

$$(4.6) \qquad \log\left[f(\boldsymbol{\eta}_T)\right] = -(T/2)\log(2\pi) - \tfrac{1}{2}\log(|\Gamma|) - \tfrac{1}{2}\boldsymbol{\eta}_T'\Gamma^{-1}\boldsymbol{\eta}_T.$$

If η were white noise, Γ would be a diagonal matrix, and conventional maximization of $\log\left[f(\boldsymbol{\eta}_T)\right]$ with respect to $\mathbf{a}$ would yield best linear unbiased estimates. These, of course, would be identical to unweighted linear least-squares estimates. In this case Γ does not enter the solution at all. Often η is an autocorrelated sequence, and Γ is required for the maximum likelihood estimates of $\mathbf{a}$. The same holds true for Aitken's generalized least-squares procedure, which minimizes $\boldsymbol{\eta}_T'\Gamma^{-1}\boldsymbol{\eta}_T$.

Since Γ is generally unknown, the maximum likelihood estimates cannot be directly computed. Several methods of circumventing the problem of an unknown covariance matrix are used in practice. One approach is to assume an error model for η that leads to a simplifying transformation of variables. For example, if η were a normal Markov process such that†

$$\eta_t = \alpha\eta_{t-1} + \epsilon_t, \qquad 0 < |\alpha| < 1,$$

ϵ being white noise, then

$$(4.7) \qquad Y_t = \alpha Y_{t-1} + \sum_{s=-r_1}^{r_2} b_s X_{t-s} + \epsilon_t,$$

$$b_s = \begin{cases} a_s & s = -r_1 \\ a_s - \alpha a_{s-1} & -r_1 < s < r_2 \\ -\alpha a_s & s = r_2 \end{cases}.$$

Using Expression (4.7) one may estimate α and $\mathbf{b}$ by the method of maximum likelihood, or equivalently by linear least-squares, and

† See Malinvaud [62, p. 485].

then solve for **a**. There remains, of course, the question of how adequate a representation the Markov error model is for explaining autocorrelation in η.

As an alternative approach one may use the unweighted linear least-squares method to estimate **a**, compute the residuals from the regression line and use the residuals to estimate Γ. The resulting covariance matrix estimate is statistically consistent, provided that r_1 and r_2 are correctly specified in the initial regression analysis. Replacing Γ by its estimate in (4.6), one may then solve for an estimate of **a** that is more efficient than the estimate originally computed.

The fact that we can estimate g_η consistently, as described in Chap. 2, suggests yet another approach. Since

$$R_{\eta,\tau} = \int_{-\pi}^{\pi} g_\eta(\lambda)e^{i\lambda\tau}\, d\lambda,$$

it seems reasonable to estimate $R_{\eta,\tau}$ by

$$\hat{R}_{\eta,\tau} = (2m)^{-1} \sum_{j=-m+1}^{m} \hat{g}_\eta(\lambda_j)e^{i\lambda_j\tau} \qquad \lambda_j = \pi j/m$$

where $2\pi/m$ is the frequency interval between spectrum estimates. Substituting these estimates into Γ one can solve (4.6) for an estimate of **a** that is again more efficient than the unweighted linear least-squares solution. Notice that, in contrast to the two-stage procedure described above, the present approach for estimating Γ is independent of the specification of r_1 and r_2. This property offers a distinct advantage when r_1 and r_2 are also unknown a priori.

This approach with some modification is the essence of Hannan's procedure for estimating coefficients in distributed lag models. The remainder of this chapter describes the necessary formalism that will enable the reader to understand the spectral approach in considerably more detail than just discussed.

The spectral approach makes use of certain properties of the covariance matrix Γ. Notice that all the elements on a given diagonal are identical and in addition the off-diagonals are symmetric around the main diagonal. A matrix with these properties is called a Toeplitz form and has a property which is very convenient for our purposes [33, p. 112]. Consider the $T \times T$ matrix **U** with the element $T^{-\frac{1}{2}}e^{i\lambda_h(k-n)}$

$$\lambda_h = \pi(h - n)/n, \qquad n = [T/2]$$

in the row h column k position. Denoting the transposed conjugate of $\mathbf{U}$ by $\mathbf{U}^*$, the elements of the matrix

$$\mathbf{V} = \mathbf{U}\boldsymbol{\Gamma}\mathbf{U}^*$$

are

$$v_{hl} = T^{-1} \sum_{j,k=-n+1}^{n} R_{\eta,k-j} e^{i(\lambda_l j - \lambda_h k)}.$$

Replacing $R_{\eta,k-j}$ by its spectral representation leads to

$$v_{hl} = \int_{-\pi}^{\pi} g_{\eta}(\lambda) e^{i(\lambda_l - \lambda_h)/2} \left[\frac{\sin n(\lambda_l - \lambda) \sin n(\lambda_h - \lambda)}{T \sin \frac{1}{2}(\lambda_l - \lambda) \sin \frac{1}{2}(\lambda_h - \lambda)} \right] d\lambda.$$

Notice that as T becomes large, the quantity in brackets converges to zero everywhere in $(-\pi, \pi)$ except at $\lambda_h = \lambda_l = \lambda$. By an appropriate limiting argument, one may show that

$$\lim_{T \to \infty} v_{hl} = \begin{cases} 2\pi g_{\eta}(\lambda_h) & h = l \\ 0 & h \neq l, \end{cases}$$

so that $2\pi\mathbf{V}$ is asymptotically a diagonal matrix whose h^{th} element is the spectrum ordinate $g_{\eta}(\lambda_h)$.

The matrix $\mathbf{U}$ is unitary, for the elements of $\mathbf{U}\mathbf{U}^*$ are

$$T^{-1} e^{i(\lambda_l - \lambda_h)n} \sum_{k=1}^{2n} e^{i(\lambda_h - \lambda_l)k} = \begin{cases} 1 & h = l \\ 0 & h \neq l. \end{cases}$$

so that

$$\mathbf{U}\mathbf{U}^* = \mathbf{I}$$

$$\mathbf{U}^{-1} = \mathbf{U}^*,$$

$\mathbf{I}$ being the identity matrix. Then we may also write asymptotically

$$\mathbf{V}^{-1} = (\mathbf{U}\boldsymbol{\Gamma}\mathbf{U}^*)^{-1} = \mathbf{U}\boldsymbol{\Gamma}^{-1}\mathbf{U}^*$$

$$\boldsymbol{\Gamma}^{-1} = \mathbf{U}^*\mathbf{V}^{-1}\mathbf{U}.$$

We see that for sufficiently large T we may regard $\mathbf{V}^{-1}$ as a diagonal matrix whose elements we can consistently estimate by $\hat{g}_{\eta}^{-1}(\lambda_h)$. Hereafter we assume T to be sufficiently large.

The exponent of the likelihood function may now be written as

$$(4.8a) \qquad \eta_T' \boldsymbol{\Gamma}^{-1} \eta_T = \eta_T' \mathbf{U}^* \mathbf{V}^{-1} \mathbf{U} \eta_T = (4\pi)^{-1} \sum_{j=-n+1}^{n} g_{\eta}^{-1}(\lambda_j) I_{\eta}(\lambda_j),$$

$$I_{\eta}(\lambda_j) = (2/T) \left| \sum_{t=-n+1}^{n} \eta_{t+n} e^{-i\lambda_j t} \right|^2, \qquad \lambda_j = \pi j/n.$$

Substituting Expression (4.3a), we have

$$(4.8b) \quad \eta_T' \mathbf{U}^* \mathbf{V}^{-1} \mathbf{U} \eta_T = (4\pi)^{-1} \sum_{j=-n+1}^{n} g_\eta^{-1}(\lambda_j) \bigg[I_y(\lambda_j)$$

$$+ \sum_{k,s=-r_1}^{r_2} a_k a_s I_x(\lambda_j) e^{i\lambda_j(s-k)} - 2 \sum_{s=-r_1}^{r_2} a_s I_{xy}(\lambda_j) e^{i\lambda_j s} \bigg],$$

since

$$\text{Im} \, [I_{xy}(\lambda)] = -\text{Im} \, [I_{xy}(-\lambda)],$$

where Im [ ] denotes the imaginary part of the quantity in brackets. Maximizing (4.8b) with respect to **a** leads to $r_1 + r_2 + 1$ simultaneous equations of the form

$$(4.9) \quad \sum_{s=-r_1}^{r_2} \tilde{a}_s \sum_{j=-n+1}^{n} g_\eta^{-1}(\lambda_j) I_x(\lambda_j) e^{i\lambda_j(l-s)}$$

$$= \sum_{j=-n+1}^{n} g_\eta^{-1}(\lambda_j) I_{xy}(\lambda_j) e^{i\lambda_j l},$$

$$l = -r_1, -r_1 + 1, \ldots, 0, \ldots, r_2 - 1, r_2,$$

where the tilde over a_s denotes an estimator.

We now come to a point of decision regarding the use of periodogram ordinates in estimating **a**. As discussed in Chap. 3, periodogram ordinates asympotically have correct expectations but are not consistent. Replacing them by their corresponding spectrum estimates, which are consistent, improves the reliability of the sample averages in (4.9) which, in turn, leads to an improved estimate of **a**.

Another consideration also plays a role. We note that the weighted summation of ordinates is taken over the interval $(-\pi, \pi)$. All that is actually required is that these ordinates give a representative picture of the spectrum. The spacing π/n is considerably smaller than the estimator's bandwidth $K_M^{-1}(0)$, so that, if the spectrum is relatively constant over the bandwidth, we would not expect much variation between adjacent sample spectrum ordinates. It is therefore possible to find an m smaller than n, such that little sacrifice of detail occurs by reducing the density of points in the summation from $2n$ to $2m$. We also make the substitutions described in Point 4.

The final form of the estimation equations is

$$(4.10a) \qquad\qquad \tilde{\mathbf{a}} \hat{\mathbf{H}} = \hat{\mathbf{d}},$$

where the elements of $\hat{\mathbf{H}}$ and $\hat{\mathbf{c}}$ are

(4.10b) $\qquad \hat{h}_{ls} = (2m)^{-1} \sum_{j=-m+1}^{m} g_\eta^{-1}(\lambda_j)\hat{g}_x(\lambda_j)e^{i\lambda_j(l-s)},$

(4.10c) $\qquad \hat{d}_l = (2m)^{-1} \sum_{j=-m+1}^{m} g_\eta^{-1}(\lambda_j)\hat{g}_{xy}(\lambda_j)e^{i\lambda_j l},$

$$\lambda_j = \pi j/m,$$

respectively.

These results follow from minimizing the expression $\eta_T' \Gamma^{-1} \eta_T$ for large T. This is the same expression that one minimizes using Aitken's generalized least-squares method [2], where normality is not required.

The coefficients in Expression (4.10a) are weighted by the quantity $\hat{g}_x(\lambda)/g_\eta(\lambda)$ summed over all frequency bands. Noting that

(4.11a) $\qquad\qquad g_\eta(\lambda) = g_y(\lambda)[1 - \gamma_{xy}^2(\lambda)],$

we may write the *signal-to-noise* ratio as

(4.11b) $\qquad g_x(\lambda)/g_\eta(\lambda) = g_x(\lambda)/\{g_y(\lambda)[1 - \gamma_{xy}^2(\lambda)]\}.$

The weighting of the coefficients is then a function of the coherence $\gamma_{xy}(\lambda)$ and the ratio of the spectra $\hat{g}_x(\lambda)/g_y(\lambda)$ in all frequency bands.

Figure 26 shows the sample signal-to-noise ratio for shipments and inventories. We note that the ratio varies by orders of magnitude over the frequency intervals, being small at $\pi/6$ and large at the remaining seasonals. In Section 4.5 we shall describe how the signal-to-noise ratio plays an important role in simplifying our estimation procedure.

The residual spectrum g_η is unknown. In contrast to the other approaches described in Section 4.2, however, one may derive a consistent estimator of g_η without the necessity of iterating the solution. For simplicity of exposition we do not make this substitution until Section 4.6.

4.3. The Sampling Properties of $\tilde{\mathbf{a}}$

By substituting (4.4) into (4.10c) and rearranging terms we have

$$(\tilde{\mathbf{a}} - \mathbf{a})\hat{\mathbf{H}} = \hat{\mathbf{c}}$$

where

$$\hat{c}_l = (2m)^{-1} \sum_{j=-m+1}^{m} g_\eta^{-1}(\lambda_j)\hat{g}_{x\eta}(\lambda_j)e^{i\lambda_j l}.$$

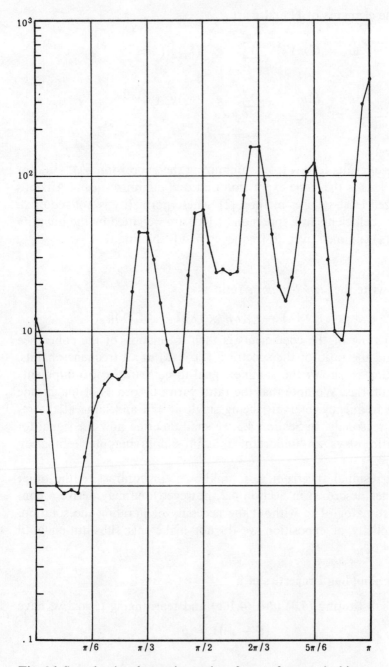

Fig. 26 Sample signal-to-noise ratio of manufacturers' shipments and inventories of durable goods, seasonally unadjusted

For large T and M we have

$$E[\hat{g}_{x\eta}(\lambda)] \sim g_{x\eta}(\lambda) = 0,$$

so that $\hat{c}_l$ asymptotically has a zero mean. Then $\tilde{a}$ is an asymptotically unbiased estimator of **a**.

The elements of the covariance matrix of $\hat{c}$ are

$$\text{cov}\,(\hat{c}_l, \hat{c}_s) = (2m)^{-2} \sum_{j,h=-m+1}^{M} g_\eta^{-1}(\lambda_j)g_\eta^{-1}(\lambda_h)$$
$$\times \text{cov}\,[\hat{g}_{x\eta}(\lambda_j)\hat{g}_{x\eta}(\lambda_h)]e^{i(\lambda_j l+\lambda_h s)}.$$

From Expression (3.29b), we have

$$\lim_{T\to\infty} T\,\text{cov}\,[\hat{g}_{x\eta}(\lambda_j)\hat{g}_{x\eta}(\lambda_h)]$$
$$= 2\pi \int_{-\pi}^{\pi} g_x(\omega)g_\eta(\omega)K_M(\omega + \lambda_h)K_M(\omega - \lambda_j)\,d\omega,$$

so that

$$\lim_{T\to\infty} T\,\text{cov}\,(\hat{c}_l, \hat{c}_s) = 2\pi \int_{-\pi}^{\pi} g_\eta(\omega)g_x(\omega)\bigg[(2m)^{-2} \sum_{j,h=-m+1}^{m} g_\eta^{-1}(\lambda_j)$$
$$\times K_M(\omega + \lambda_j)g_\eta^{-1}(\lambda_h)K_M(\omega + \lambda_h)e^{i(\lambda_j l+\lambda_h s)}\bigg]\,d\omega.$$

As the density of points in $(-\pi, \pi)$ is increased, we have

$$\lim_{m\to\infty} (2m)^{-1} \sum_{j=-m+1}^{m} g_\eta^{-1}(\lambda_j)K_M(\omega - \lambda_j)e^{i\lambda_j l}$$
$$= (2\pi)^{-1} \int_{-\pi}^{\pi} g_\eta^{-1}(\lambda)K_M(\omega - \lambda)e^{i\lambda l}\,d\lambda.$$

Now increasing the number of lags M leads to

$$\lim_{M\to\infty} \int_{-\pi}^{\pi} g_\eta^{-1}(\lambda)K_M(\omega - \lambda)e^{i\lambda l}\,d\lambda = g_\eta^{-1}(\omega)e^{i\omega l},$$

so that the terms of the asymptotic covariance matrix of $\hat{c}$ are approximately

$$(4.12) \qquad (2\pi T)^{-1} \int_{-\pi}^{\pi} g_\eta^{-1}(\omega)g_x(\omega)e^{i\omega(l-s)}\,d\omega.$$

Since cross-spectrum estimates are asymptotically normal, it follows that for sufficiently large T, M and m, $\hat{c}$ is approximately normally distributed with zero mean vector and covariance matrix, with elements given by Expression (4.12).

Now, concentrating on $\hat{\mathbf{H}}$, we may write

$$\hat{h}_{ls} = (4\pi m)^{-1} \sum_{j=-m+1}^{m} g_{\eta}^{-1}(\lambda_j) \sum_{\tau=-M}^{M} k_{M,\tau} C_{x,\tau} e^{i\lambda_j(l-s-\tau)}$$

by substituting the spectrum estimator into Expression (4.10b). Since g_{η} is positive, it follows from a theorem due to Wiener that $g_{\eta}^{-1}(\lambda)$ has an absolutely convergent Fourier series such that[†]

$$g_{\eta}^{-1}(\lambda) = (2\pi)^{-1} \sum_{\nu=-\infty}^{\infty} \mu_{\nu} e^{-i\lambda\nu}, \qquad \sum_{\nu=-\infty}^{\infty} |\mu_{\nu}| < \infty.$$

Then

$$\hat{h}_{sl} = (2\pi)^{-2} \sum_{\nu=-\infty}^{\infty} \mu_{\nu} \sum_{\tau=-M}^{M} k_{M,\tau} C_{x,\tau} \left[(2m)^{-1} \sum_{j=-m+1}^{m} e^{i\lambda_j(l-s-\tau-\nu)} \right]$$

$$= (2\pi)^{-2} \sum_{\tau=-M}^{M} \mu_{l-s-\tau} k_{M,\tau} C_{x,\tau}.$$

The mean-square convergence of $C_{x,\tau}$ has been shown in Section 3.2. We may then replace $C_{x,\tau}$ by its limit in probability $R_{x,\tau}$ so that

$$p \lim_{T \to \infty} \hat{h}_{sl} = (2\pi)^{-2} \sum_{\tau=-M}^{M} \mu_{l-s-\tau} k_{M,\tau} R_{x,\tau}.$$

Since, in addition,

$$\lim_{M \to \infty} k_{M,\nu} = 1$$

for all ν, $\hat{h}_{sl}$ converges in probability to

$$(4.13) \qquad h_{sl} = (2\pi)^{-1} \int_{-\pi}^{\pi} g_{\eta}^{-1}(\omega) g_x(\lambda) e^{i\omega(l-s)} \, d\omega.$$

If $\mathbf{H}$ with elements as in Expression (4.13) is the limit in probability of $\hat{\mathbf{H}}$, then it follows from a theorem due to Slutzky that $T^{\frac{1}{2}}(\tilde{\mathbf{a}} - \mathbf{a})$ has the same limiting distribution as $T^{\frac{1}{2}}\mathbf{H}^{-1}\hat{\mathbf{c}}$.[‡] Noting that the asymptotic covariance matrix of $T^{\frac{1}{2}}\hat{\mathbf{c}}$ is $\mathbf{H}$, we may regard $(\tilde{\mathbf{a}} - \mathbf{a})$ as having a multivariate normal distribution with mean vector zero and covariance matrix given approximately by $T^{-1}\mathbf{H}^{-1}$. This is the covariance matrix of the best linear unbiased estimator as derived by Grenander and Rosenblatt [32, Chap. 7]. As a consequence, the

---

[†] See Doob [20, pp. 159–160].
[‡] See Cramér [16, pp. 254–255].

estimator $\tilde{\mathbf{a}}$ is asymptotically efficient. As our estimator of the covariance matrix we use $T^{-1}\hat{\mathbf{H}}^{-1}$.

As Malinvaud mentions in his book [62, p. 482] the multicollinearity between lagged variables unfortunately inflates the covariance matrix of the estimated coefficients. In our terminology, the autocorrelation in the X sequence would cause this inflation if η were a white noise process. The covariances between estimates would in fact increase in magnitude as the autocorrelation in X increases. Inspection of (4.13) confirms this.

One might expect that prewhitening the X and Y time series, which reduces autocorrelation but preserves coherence, would reduce the covariance matrix. If both series are prewhitened by the same filter, then the ratio $g_x(\lambda)/g_\eta(\lambda)$ is unaffected and, consequently, so is the covariance matrix. The conjecture is therefore untrue. Here prewhitening enables us to get spectrum estimates that are closer to the population spectra; however, prewhitening does not assist us in reducing the covariance matrix.

Examining (4.13) we note that, if the signal-to-noise ratio is relatively constant over $(-\pi, \pi)$, then the off-diagonal terms are close to zero and the estimated coefficients are relatively independent. We therefore have the interesting result that in a distributed lag model there is an advantage in having an autocorrelated residual process whose spectrum closely approximates that of the X process. Hence concern about autocorrelation in X is somewhat mitigated. In Section 4.5 we shall investigate the constant signal-to-noise ratio case in more detail.

4.4. The Uniform Residual Spectrum

Suppose η is white noise and, therefore, has a uniform spectrum g_η. Expression (4.8) reduces to

$$(4.14) \qquad (2m)^{-1} \sum_{s=-r_1}^{r_2} \tilde{a}_s \sum_{j=-m+1}^{m} \hat{g}_x(\lambda_j) e^{i\lambda_j(l-s)}$$

$$= (2m)^{-1} \sum_{j=-m+1}^{m} g_{xy}(\lambda_j) e^{i\lambda_j l},$$

$$(2\pi)^{-1} \sum_{s=-r_1}^{r_2} \tilde{a}_s \sum_{\tau=-M}^{M} k_{M,\tau} C_{x,\tau} \left[(2m)^{-1} \sum_{j=-m+1}^{m} e^{i\lambda_j(l-s-\tau)} \right]$$

$$= (2\pi)^{-1} \sum_{\tau=-M}^{M} k_{M,\tau} C_{xy,\tau} \left[(2m)^{-1} \sum_{j=-m+1}^{m} e^{i\lambda_j(l-\tau)} \right],$$

$$\sum_{s=-r_1}^{r_2} \tilde{a}_s k_{M,l-s} C_{x,l-s} = k_{M,l} C_{xy,l}.$$

If M is large enough, the weighting function k is close to unity so that

$$\sum_{s=-r_1}^{r_2} \tilde{a}_s C_{x,s-l} \sim C_{xy,l},$$

which is the unweighted least-squares estimator. This is not surprising, for we note that white noise implies Γ is a diagonal matrix with $R_{\eta,0}$ in all positions on the main diagonal. The inverse of the asymptotic covariance matrix of $T^{\frac{1}{2}}\tilde{\mathbf{a}}$ then has elements

$$(2\pi)^{-1} \int_{-\pi}^{\pi} g_\eta^{-1}(\omega) g_x(\omega) e^{i\omega(l-s)} \, d\omega = R_{x,l-s}/R_{\eta,0},$$

since

$$2\pi g_\eta(\omega) = R_{\eta,0}.$$

Notice that, if η is white noise, the problem of estimating g_η disappears. This is a convenient result and it suggests that an inspection of

(4.15) $$\hat{g}_\eta(\lambda) = \hat{g}_y(\lambda)[1 - \hat{\gamma}^2(\lambda)]$$

be made prior to estimating **a**. If the estimated residual spectrum is reasonably flat, then assuming g_η to be uniform is probably not far from the truth. In some cases one may be able to prewhiten the time series on Y so that the residual spectrum is flat. If the same filter is applied to X, then one may use the unweighted linear least-squares method on the two prewhitened time series to estimate the coefficients of the original distributed lag model.†

4.5. Constant Signal-to-Noise Ratio

If $g_\eta^{-1}(\lambda) g_x(\lambda)$, the signal-to-noise ratio at frequency λ, is constant for all λ in $(-\pi, \pi)$, then we may write the asymptotically efficient estimator as

(4.16) $$\tilde{a}_l = \sum_{s=-r_1}^{r_2} \tilde{a}_s \left[(2m)^{-1} \sum_{j=-m+1}^{m} e^{i\lambda_j(l-s)} \right]$$

$$= (2m)^{-1} \sum_{j=-m+1}^{m} g_x^{-1}(\lambda_j) \hat{g}_{xy}(\lambda_j) e^{i\lambda_j l}.$$

† See Section 2.28.

This formulation has a considerable advantage for it enables us to estimate successive a_l's without recomputing earlier ones. This is desirable, particularly when r_1 and r_2 are not known in advance and there is little evidence on which to make an informed guess.

Substituting (4.4) into (4.16) we have asymptotically

$$E(\tilde{a}_l) = a_l.$$

The covariance matrix of $\tilde{a}$ has elements

$$\text{cov}\,(\tilde{a}_l, \tilde{a}_s) = (2m)^{-1} \sum_{j,h=-m+1}^{m} g_x^{-1}(\lambda_j) g_x^{-1}(\lambda_h)$$
$$\times E[\hat{g}_{x\eta}(\lambda_j)\hat{g}_{x\eta}(\lambda_h)] e^{i(\lambda_j l + \lambda_h s)}$$

so that, as before,

(4.17a) $$\lim_{M\to\infty} \lim_{m\to\infty} \lim_{T\to\infty} T \,\text{cov}\,(\tilde{a}_l, \tilde{a}_s)$$
$$= (2\pi)^{-1} \int_{-\pi}^{\pi} g_x^{-1}(\omega) g_\eta(\omega) e^{i\omega(l-s)} \, d\omega,$$

which equals the signal-to-noise ratio when $l = s$ and vanishes elsewhere. Therefore, the elements of $\tilde{a}$ are asymptotically independent. Inspection of a graph of the sample SNR for all frequencies considered may consequently lead to a considerable simplification of the estimation problem. Looking at Fig. 26, we note the large variations in the ratio, thereby precluding the constant signal-to-noise ratio assumption.

The simplicity of the estimator (4.16) suggests that we may want to use it even though the signal-to-noise ratio is not constant. Expression (4.17a) still defines the asymptotic covariance matrix but the covariances no longer vanish. As an estimate of cov $(\tilde{a}_l, \tilde{a}_s)$ we use

(4.17b) $$(2mT)^{-1} \sum_{j=-m+1}^{m} \hat{g}_x^{-1}(\lambda_j) \hat{g}_\eta(\lambda_j) e^{i\lambda_j(l-s)}.$$

To measure efficiency, we may examine the ratio of the determinants of the sample covariance matrices for the estimators (4.16) and (4.10a). The sample covariance matrix of (4.10a) is the inverse of the matrix with terms given in (4.10b).

The appeal here is that we may determine the efficiency of (4.16) in advance. This advantage is considerably enhanced in the multivariate case, where a favorable evaluation of efficiency may considerably reduce the dimensionality of the set of equations to be solved. This case is discussed in Section 4.15.

4.6. The Substitution of Consistent Estimators for g_η and g_x

From the approximation in the assumption in Point 6, we have

$$\hat{g}_\eta^{-1}(\lambda)\hat{g}_{x\eta}(\lambda) = g_\eta^{-1}(\lambda)\hat{g}_{x\eta}(\lambda)$$

so that, to order (M/T), the covariance matrix of $\hat{\mathbf{c}}$, with a consistent estimator replacing g_η, is asymptotically the same. If (T/M) is large enough, we may safely ignore the higher-order variance contributions.

Turning to $\hat{\mathbf{H}}$, we note that

$$\hat{g}_\eta^{-1}(\lambda)\hat{g}_x(\lambda) \sim g_\eta^{-1}(\lambda)g_x(\lambda) + g_\eta^{-1}(\lambda)[\hat{g}_x(\lambda) - g_x(\lambda)]$$
$$- g_\eta^{-2}(\lambda)g_x(\lambda)[\hat{g}_\eta(\lambda) - g_\eta(\lambda)],$$

which for well-resolved estimates has to order (M/T)

$$\text{var}\,[\hat{g}_\eta^{-1}(\lambda)\hat{g}_x(\lambda)] \sim 2\delta_\lambda \Psi_{T,M} g_\eta^{-2}(\lambda)g_x^2(\lambda).$$

Since

$$\text{var}\,[g_\eta^{-1}(\lambda)\hat{g}_x(\lambda)] \sim \delta_\lambda \Psi_{T,M} g_\eta^{-2}(\lambda)g_x^2(\lambda),$$

the substitution of a consistent estimator for g_η does not invalidate the basic results but it does slow down convergence. The most immediate choice of an estimator for g_η is Expression (4.15), which is easily computed from estimates of g_y and γ_{xy}.

In the constant signal-to-noise ratio case, Expression (4.16) gives the appropriate estimator. Since g_x is generally unknown, we replace it by its estimator so that

(4.18) $$\tilde{a}_l = (2m)^{-1} \sum_{j=-m+1}^{m} \hat{g}_x^{-1}(\lambda_j)\hat{g}_{xy}(\lambda_j)e^{i\lambda_j l}.$$

Using approximation in the assumption of Point 6, one may show that to order T^{-1} the asymptotic covariance matrix is given by Expression (4.17). Notice that the problem of convergence on the left side of the estimation equations does not enter the problem.

4.7. The Choice of m

As stated earlier, m should be chosen to give an adequate representation of the spectral structure. It is reasonable to assume that the spectra of interest are reasonably flat over intervals of $2\pi/M$ radians, since the bandwidths of the Tukey-Hanning and Parzen windows are $2\pi/M$ and $2.6\pi/M$, respectively. Setting m equal to $M/2$ should therefore result in a good description of the spectra. In addition, this choice has the advantage of reducing the number of design parameters to be considered.

The covariance matrix may generally be estimated by $T^{-1}\hat{\mathbf{H}}^{-1}$, which leads to the correct limiting form. If T and M are not sufficiently large, then the use of this estimator becomes somewhat tenuous. In this case the spectrum will vary over the bandwidth and estimates will not be well resolved.

4.8. Adjustment for Nonzero Means

In general we work with time series having nonzero means. Here one must consider the expression

$$Y_t = b + \sum_{s=-r_1}^{r_2} a_s X_{t-s} + \eta_t.$$

To estimate b, we use

$$\tilde{b} = \overline{Y} - \overline{X} \sum_{s=-r_1}^{r_2} \tilde{a}_s,$$

where $\overline{X}$ and $\overline{Y}$ are the sample means of the X and Y time series, respectively—not their prewhitened means.

4.9. The Adaptive Model

Very often certain restrictions are imposed on **a** as a consequence of economic theory and a desire for statistical simplicity. In a number of cases, the adaptive model†

(4.19) $a_s = \alpha\beta^s$ $s = 0, 1, \ldots, \infty$ $|\beta| < 1$

satisfies these requirements, for we note that it implies

$$Y_t = \alpha X_t + \beta Y_{t-1} + \eta_t - \beta\eta_{t-1}.$$

In this way it is possible to account for current influences in X and past influences in Y.

This specification is easily handled for

(4.20) $A(\lambda) = \alpha^{-1} \sum_{s=0}^{\infty} a_s e^{-i\lambda s} = (1 - \beta^{-i\lambda})^{-1},$

---

† See Nerlove [66].

so that the logarithm of the likelihood function, ignoring the constants, is approximately

$$(2m)^{-1} \sum_{j=-m+1}^{m} \{g_\eta^{-1}(\lambda_j)[\hat{g}_y(\lambda_j) + \alpha^2|A(\lambda)|^2\hat{g}_x(\lambda_j)$$
$$- \alpha A^*(\lambda_j)\hat{g}_{xy}(\lambda_j) - \alpha A(\lambda_j)\hat{g}_{xy}^*(\lambda_j)]\}.$$

Maximization with respect to α and β leads to:

$$(4.21) \qquad (2m)^{-1} \sum_{j=-m+1}^{m} \tilde{g}_w^{-1}(\lambda_j)\hat{g}_{xy}(\lambda_j)\begin{bmatrix} 1 \\ \tilde{A}^*(\lambda_j)e^{i\lambda_j} \end{bmatrix} = \hat{\mathbf{H}}\begin{bmatrix} \tilde{\alpha} \\ \tilde{\beta} \end{bmatrix},$$

$$\hat{\mathbf{H}} = (2m)^{-1} \sum_{j=-m+1}^{m} \tilde{g}_w^{-1}(\lambda_j)\begin{bmatrix} \hat{g}_x(\lambda_j) & \hat{g}_{xy}(\lambda_j)e^{-i\lambda_j} \\ \tilde{\alpha}\tilde{A}^*(\lambda_j)\hat{g}_x(\lambda_j)e^{i\lambda_j} & \tilde{A}(\lambda_j)\hat{g}_{xy}(\lambda_j) \end{bmatrix},$$

$$\tilde{g}_w(\lambda) = g_\eta(\lambda)/|\tilde{A}(\lambda)|^2.$$

Unfortunately, these equations are nonlinear in $\tilde{\beta}$. They may be solved iteratively, however, the recursive estimate of $\tilde{\beta}$ being substituted into the sample function $\tilde{A}$ until convergence occurs. As before, a consistent estimator of g_η is also necessary.

To derive the sampling properties of $\tilde{\alpha}$ and $\tilde{\beta}$, it is convenient to regard the frequency response function A as known. Then noting that approximately

$$\hat{g}_{xy}(\lambda_j) = \alpha\hat{g}_x(\lambda_j) + \beta\hat{g}_{xy}(\lambda_j)e^{-i\lambda_j} + \hat{g}_{x\eta}(\lambda_j)/A(\lambda_j),$$

one may write

$$(4.22) \qquad\qquad \hat{\mathbf{H}}\begin{pmatrix} \tilde{\alpha} - \alpha \\ \tilde{\beta} - \beta \end{pmatrix} = \hat{\mathbf{c}},$$

$$\hat{\mathbf{c}} = (2m)^{-1} \sum_{j=-m+1}^{m} g_w^{-1}(\lambda_j)\begin{bmatrix} \hat{g}_{xw}(\lambda_j) \\ A^*(\lambda_j)\hat{g}_{xw}(\lambda_j)e^{i\lambda_j} \end{bmatrix},$$

$$\hat{g}_{xw}(\lambda_j) = \hat{g}_{x\eta}(\lambda_j)/A(\lambda_j).$$

As before, $T^{\frac{1}{2}}\hat{\mathbf{c}}$ is asymptotically normal.

Since the cross spectrum $g_{x\eta}$ is zero, the estimators $\tilde{\alpha}$ and $\tilde{\beta}$ are asymptotically unbiased. Using the method of Section 4.3, one may show the asymptotic covariance matrix of $T^{\frac{1}{2}}\hat{\mathbf{c}}$ to be

$$(4.23) \qquad\qquad 2\pi \int_{-\pi}^{\pi} g_w^{-1}(\lambda)g_x(\lambda)\mathbf{A}(\lambda)\,d\lambda.$$

$$\mathbf{A}(\lambda) = \begin{bmatrix} 1 & A(\lambda)e^{-i\lambda} \\ A^*(\lambda)e^{i\lambda} & |A(\lambda)|^2 \end{bmatrix}.$$

Noting that the matrix $\hat{\mathbf{H}}$ can be written as

$$\hat{\mathbf{H}} = (2m)^{-1} \sum_{j=-m+1}^{m} g_w^{-1}(\lambda_j)$$

$$\times \begin{bmatrix} \hat{g}_x(\lambda_j) & [\alpha A(\lambda_j)\hat{g}_x(\lambda_j) + \hat{g}_{x\eta}(\lambda_j)]e^{-i\lambda_j} \\ \alpha A^*(\lambda_j)\hat{g}_x(\lambda_j)e^{i\lambda_j} & A^*(\lambda_j)[\alpha A(\lambda_j)\hat{g}_x(\lambda_j) + \hat{g}_{x\eta}(\lambda_j)] \end{bmatrix},$$

one may show that $\hat{\mathbf{H}}$ converges in probability to

(4.24) $$\mathbf{H} = 2\pi \int_{-\pi}^{\pi} g_w^{-1}(\lambda)g_x(\lambda)A(\lambda)\begin{bmatrix} 1 & 0 \\ 0 & \alpha \end{bmatrix} d\lambda.$$

As a result, the asymptotic covariance matrix of $\tilde{\alpha}$ and $\tilde{\beta}$ is approximately

(4.25) $$T^{-1}\left\{\mathbf{H}\begin{bmatrix} 1 & 0 \\ 0 & \alpha \end{bmatrix}\right\}^{-1} = T^{-1}\begin{bmatrix} 1 & 0 \\ 0 & \alpha^{-1} \end{bmatrix}\mathbf{H}^{-1}.$$

These results continue to hold when g_w and A are replaced in $\hat{\mathbf{H}}$ by consistent estimators. For the estimator of the covariance matrix (4.25), we replace α and $\mathbf{H}$ by their respective estimates. It is again convenient to set m equal to $M/2$, bearing in mind the discussion of Section 4.7.

The Hannan estimator of (α, β) has advantages over other estimators. Hannan [42] has shown it to be more efficient than the estimator suggested by Leviatan [59], and Amemiya and Fuller [2] have shown it to be asymptotically more efficient than the estimator proposed by Klein [55]. The Leviatan and Klein estimators may be used to derive initial estimates for the iterative process.

To adjust for the nonzero mean in X and Y we consider the specification

$$Y_t = b + \alpha \sum_{s=0}^{\infty} \beta^s X_{t-s} + \eta_t.$$

Then as our estimate of b we take

$$\tilde{b} = \overline{Y} - \tilde{\alpha}\overline{X}/(1 - \tilde{\beta}).$$

4.10. Testing the Validity of a Bivariate Model

Model specification is generally a trial-and-error procedure, even when based on economic theory. One may easily hypothesize a linear association between two economic phenomena, but seldom is enough prior information available to specify r_1 and r_2 correctly. A procedure is therefore desirable to test the validity of a particular model. By validity we mean the extent to which a model is consistent with the data used to estimate its coefficients.

Ideally, we would like to validate a model with data other than the original sample. Here successful validation gives us confidence in the model's universal applicability to the statistical populations considered rather than just to the original sample drawn from these populations. This testing can be accomplished by dividing time series into two halves, using the first half for estimation and the second half for testing.

Economic times series are seldom long enough to afford us this luxury, so we must be content with testing validity with the same data. This kind of testing, while a compromise, is especially important when model specification is a consequence of economic theory. If we cannot validate the model with the original sample, we cannot seriously consider the model to apply universally.

To test the validity of the model with the data, we may test the hypothesis that the coherence function $\gamma_{x\eta}$ is zero. The testing procedures are described in Section 3.19. The sample coherence is computed using the time series on X and the residual time series computed from

$$\tilde{\eta}_t = Y_t - \tilde{b} - \sum_{s=-r_1}^{r_2} \tilde{a}_s X_{t-s}.$$

The number of observations is reduced here from T to $T - r_1 - r_2$.[†] Also, plotting the sample coherence function $\hat{\gamma}_{x\eta}$ and comparing it with $\hat{\gamma}_{xy}$ enable the investigator to determine the extent to which different frequency bands are affected by the regression.

To date, no comprehensive theory for testing the significance of distributed lag coefficients has been developed using spectral methods. Hannan [33], Hamon and Hannan [35], and Wallis [96] have suggested alternatives to the approach suggested here but the present approach has the advantage of simplicity and makes use of more generally developed procedures.

In the adaptive case, we suggest testing the hypothesis that the coherence function $\gamma_{x\eta'}$ is zero where

$$\eta'_t = \eta_t - \beta \eta_{t-1}.$$

No loss of generality occurs here since the linear filtering of η preserves its linear association with X, if any exists. The residual time series is

---

† Here we ignore the fact that $\hat{\eta}_t$ is computed using estimates of **a** and b.

computed from

$$\eta'_t = Y_t - \tilde{\alpha} X_t - \tilde{\beta} Y_{t-1} - \tilde{b}(1 - \tilde{\beta}),$$

the number of observations now being $T - 1$.

If we accept the null hypothesis, we are not necessarily minimizing the residual variance due to η. The discussion in Section 2.26 notes that η may contain an autocorrelated sequence Z that is uncorrelated with X. Inclusion of Z in the regression analysis would reduce the residual variance. If X and Z are correlated, then acceptance of the null hypothesis when Z is omitted implies that our choice of r_1 and r_2 accounts for the correlation between these sequences. Were we to include Z in the regression analysis, we would probably find our original choice of r_1 and r_2 altered.

4.11. An Example of a Distributed Lag Model

As an example, we regressed the shipments (X) time series on the inventories (Y) time series. In our first regression we set

$$r_1 = 0, \qquad r_2 = 12,$$

so that present inventories were a function of present and past shipments. The sample coherence function $\hat{\gamma}_{x\eta}$ between shipments and inventory residuals was found to be about 0.8 at low frequencies and declined somewhat thereafter. The null hypothesis of zero coherence in $(0, \pi)$ was rejected at the 0.10 significance level.

We then set

$$r_1 = r_2 = 6$$

so that present inventories were associated with future, present, and past shipments. Table 10 lists the sample coefficients along with the corresponding sample standard deviations and covariance matrix. Regarding the $\tilde{a}_s$'s as normally distributed variates, we find, on testing them individually, that all but those at lags -6, -5, and 6 are significant at the 0.10 level with a one-sided test. The constant in the regression was estimated to be

$$\tilde{b} = -994.$$

Since inventories during the sample period were of order 10^5, the constant contributes little to the model.

Table 10

Results for the Distributed Lag Model for Shipments (X) and Inventories (Y)

j	$\tilde{a}_j$ (10^{-2})	Std. Dev. (10^{-2})	-6	-5	-4	-3	-2	-1	0	1	2	3	4	5	6
									Covariance Matrix (10^{-4})						
-6	0.46	1.51	2.28	1.69	1.34	0.83	0.37	0.07	-0.58	-0.82	-0.85	-0.58	-0.40	0.17	-0.50
-5	3.71	1.85		3.42	2.72	1.87	0.98	0.24	-0.53	-1.31	-1.43	-1.21	-0.70	0.02	0.17
-4	6.61	2.05			4.20	3.24	2.13	1.08	-0.04	-0.97	-1.81	-1.80	-1.50	-0.70	-0.40
-3	10.90	2.10				4.43	3.27	2.01	0.73	-0.44	-1.27	-1.96	-1.80	-1.21	-0.58
-2	15.10	2.08					4.34	3.07	1.70	0.45	-0.56	-1.27	-1.81	-1.43	-0.85
-1	27.50	2.01						4.03	2.74	1.46	0.45	-0.44	-0.97	-1.31	-0.81
0	33.50	1.97							3.88	2.74	1.70	0.73	-0.04	-0.53	-0.58
1	29.70	2.01								4.03	3.07	2.01	1.08	0.24	0.07
2	28.10	2.08									4.34	3.27	2.13	0.97	0.37
3	23.40	2.10										4.43	3.24	1.87	0.83
4	16.20	2.05											4.20	2.72	1.34
5	5.35	1.85												3.42	1.69
6	2.44	1.51													2.28

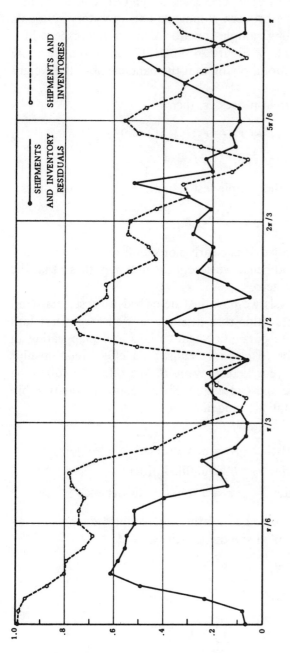

Fig. 27 Sample coherence of manufacturers' shipments and inventories of durable goods, seasonally unadjusted

Figure 27 shows the sample coherences of interest. Under the null hypothesis, the coherence between shipments and inventory residuals is zero. The corresponding sample function appears low except in a low-frequency band near $\pi/6$. Figure 26 shows this band to have a low sample signal-to-noise ratio implying that the band carries little weight in the estimation of **a**.

If X and η were normal processes, then

$$n_\lambda = 1.85(T - r_1 - r_2)/(\delta_\lambda M) \sim 9/\delta_\lambda$$

$$T = 238, \qquad M = 48.$$

Setting $L = M/2 = 24$, the sample test statistic of the null hypothesis is†

$$\Delta_{24} = -2.8.$$

Since the statistic is negative, it is clearly not significant at any reasonable significance level, and hence we accept the null hypothesis that the coherence function $\gamma_{x\eta}$ is zero.

The regression analysis we have just described was essentially an exercise in curve fitting. In this case negative as well as positive lags were needed to make the data fit. Negative lags are unappealing in economics for we prefer to consider cause and effect relationships wherein event A must occur before event B. We offer the following explanation to reconcile the cause and effect economic relationship with negative lags in a statistic model.

We define

$$X \equiv \text{shipments}, \qquad N \equiv \text{new orders},$$
$$Y \equiv \text{inventories}, \qquad U \equiv \text{unfilled orders},$$
$$P \equiv \text{production}, \qquad \epsilon \equiv \text{a random disturbance}.$$

The subscript t denotes period t and the superscript * denotes anticipated or planned levels. We have the identity

(4.26a) $$P_t + Y_{t-1} = X_t + Y_t.$$

Anticipated new orders are

(4.26b) $$N_t^* = \sum_{s=0}^{n} b_s N_{t-s},$$

---

† See Section 3.19.

and planned production is

(4.26c) $$P_t^* = U_t + N_t^* - Y_{t-1}.$$

Actual production is

(4.26d) $$P_t = P_t^* + \epsilon_t,$$

P^* and ϵ being mutually uncorrelated processes. Shipments are clearly a function of past new orders so that

(4.27a) $$X_t = \sum_{s=0}^{n} c_s N_{t-s} \qquad 0 \le c_s \le 1.$$

Later we shall add a random component to this sum.

Suppose that all new orders are shipped within two periods so that

(4.27b) $$X_t = c_0 N_t + c_1 N_{t-1} \qquad c_0 + c_1 = 1.$$

If new orders are handled on a first-in first-out basis and production time is longer than one period, it may occur that $c_1 > c_0$, so that we may express N_t as

(4.28a) $$N_t = (1/c_1) \sum_{s=0}^{\infty} (-c_0/c_1)^s X_{t+s+1}.$$

More generally, if the roots of the polynomial†

(4.28b) $$\sum_{s=0}^{n} c_s \xi^s$$

lie inside the unit circle, then we may express N_t as a linear combination of future shipments so that

(4.28c) $$N_t = \sum_{s=0}^{\infty} \delta_s X_{t+s+1},$$

where the coefficients $\{\delta_s\}$ are determined from (4.28b). Combining all these expressions yields

(4.29) $$Y_t = P_t + Y_{t-1} - X_t$$

$$= \sum_{s=0}^{n} b_s N_{t-s} - X_t + U_t + \epsilon_t$$

$$= \sum_{s=0}^{n} b_s \sum_{\tau=0}^{\infty} \delta_\tau X_{t-s+\tau+1} - X_t + U_t + \epsilon_t.$$

---

† See Section 2.24.

If we add a random disturbance ξ_t to (4.27a), $\{\xi_t\}$ and $\{N_t\}$ being mutually uncorrelated, then the terms in the double summation in (4.29) contain $X_\nu - \xi_\nu$ in place of X_ν.

This description offers an explanation of how negative lags may occur in the statistical model. In addition, it suggests that r_1 and r_2 are chosen partially to compensate for the correlation between X and ξ and possibly U. The residual process η in our model then accounts for ϵ and the part of ξ and U uncorrelated with X.

4.12. Estimation for Simultaneous Equations

The multivariate extension of the bivariate results follows directly. We consider a set of simultaneously distributed lag equations as given by (2.49). Strict stationarity and normality are assumed for all components. Under certain conditions, to be discussed later, the predetermined variables may contain lagged endogenous variables. Given a set of T observations on $\mathbf{X}$ and $\mathbf{Y}$, our purpose is to estimate the matrix $\mathbf{a}$.

The exponent of the likelihood function is now

$$\tfrac{1}{2}\boldsymbol{\eta}'_T \boldsymbol{\Gamma}^{-1} \boldsymbol{\eta}_T,$$

where the vector of residuals with pT components is

$$\boldsymbol{\eta}'_T = [\eta_{11}, \eta_{21}, \ldots, \eta_{i1}, \ldots, \eta_{p1}; \; \eta_{12}, \eta_{22}, \ldots, \eta_{i2}, \ldots, \eta_{p2};$$
$$\ldots; \; \eta_{1t}, \eta_{2t}, \ldots, \eta_{it}, \ldots, \eta_{pt};$$
$$\ldots; \; \eta_{1T}, \eta_{2T}, \ldots, \eta_{iT}, \ldots, \eta_{pT}],$$

and the covariance matrix is

$$(4.30a) \qquad \boldsymbol{\Gamma} = \begin{bmatrix} \mathbf{R}_{\eta\eta,0} & \mathbf{R}_{\eta\eta,1} & \cdots & \mathbf{R}_{\eta\eta,T-1} \\ \mathbf{R}_{\eta\eta,-1} & \mathbf{R}_{\eta\eta,0} & & \vdots \\ \vdots & & \ddots & \vdots \\ \mathbf{R}_{\eta\eta,-T+1} & \cdots & & \mathbf{R}_{\eta\eta,0} \end{bmatrix},$$

with

$$(4.30b) \qquad \mathbf{R}_{\eta\eta,\tau} = \begin{bmatrix} R_{\eta_1\eta_1,\tau} & R_{\eta_1\eta_2,\tau} & \cdots & R_{\eta_1\eta_p,\tau} \\ R_{\eta_2\eta_1,\tau} & R_{\eta_2\eta_2,\tau} & & \vdots \\ \vdots & & \ddots & \vdots \\ R_{\eta_p\eta_1,\tau} & \cdots & & R_{\eta_p\eta_p,\tau} \end{bmatrix}.$$

It can be shown that the transformation

$$(\mathbf{U} \otimes \mathbf{I})\Gamma(\mathbf{U} \otimes \mathbf{I})^*$$

asymptotically leads to

(4.31a) $\quad \mathbf{V} = \begin{bmatrix} g_\eta(\lambda_{-n+1}) & 0 & \cdots & & & & 0 \\ 0 & g_\eta(\lambda_{-n+2}) & & & & & \vdots \\ \vdots & & 0 & \ddots & & & \\ & & & & g_\eta(\lambda_0) & & \\ & & & & & \ddots & \\ 0 & & \cdots & & & & g_\eta(\lambda_n) \end{bmatrix},$

$$\lambda_j = \pi j/n, \quad n = [T/2],$$

(4.31b) $\quad g_{\eta\eta}(\lambda) = \begin{bmatrix} g_{\eta_1\eta_1}(\lambda) & g_{\eta_1\eta_2}(\lambda) & \cdots & g_{\eta_1\eta_p}(\lambda) \\ g_{\eta_2\eta_1}(\lambda) & g_{\eta_2\eta_2}(\lambda) & & \vdots \\ \vdots & & & \\ g_{\eta_p\eta_1}(\lambda) & \cdots & & g_{\eta_p\eta_p}(\lambda) \end{bmatrix},$

where $\mathbf{U} \otimes \mathbf{I}$ is a Kronecker product and $(\mathbf{U} \otimes \mathbf{I})^*$ is its complex conjugate [34, pp. 95–96].

Since the inverse of a unitary matrix is its complex conjugate, we have

(4.32a) $\quad [(\mathbf{U} \otimes \mathbf{I})\Gamma(\mathbf{U} \otimes \mathbf{I})^*]^{-1} = (\mathbf{U} \otimes \mathbf{I})\Gamma^{-1}(\mathbf{U} \otimes \mathbf{I})^*,$

the mathematical limit of (4.32a) thus being

(4.32b)

$$\mathbf{V}^{-1} = \begin{bmatrix} g_{\eta\eta}^{-1}(\lambda_{-n+1}) & 0 & \cdots & & & & 0 \\ 0 & g_{\eta\eta}^{-1}(\lambda_{-n+2}) & & & & & \vdots \\ \vdots & & & \ddots & & & \\ & & & & g_{\eta\eta}^{-1}(\lambda_0) & & \\ & & & & & \ddots & \\ 0 & & \cdots & & & & g_{\eta\eta}^{-1}(\lambda_n) \end{bmatrix}.$$

Denoting an element of the inverse matrix $g_{\eta\eta}^{-1}(\lambda)$ by $g_{\eta\eta}^{kl}(\lambda)$, we may now write the exponent of the likelihood function as approximately

(4.33a) $\qquad \eta_T'(\mathbf{U} \otimes \mathbf{I})^* \mathbf{V}^{-1}(\mathbf{U} \otimes \mathbf{I})\eta_T$

$$= (4\pi)^{-1} \sum_{j=-n+1}^{n} \sum_{k,l=1}^{p} I_{\eta_l \eta_k}(\lambda_j) g_{\eta\eta}^{kl}(\lambda_j),$$

(4.33b) $\qquad I_{\eta_l \eta_k}(\lambda) = (1/T)\left(\sum_{t=1}^{T} \eta_{l,t} e^{i\lambda t}\right)\left(\sum_{t=1}^{T} \eta_{k,t} e^{-i\lambda t}\right),$

(4.33c) $\qquad \eta_{k,t} = Y_{k,t} - \sum_{\tau=1}^{N} \sum_{s=-r_1}^{r_2} a_{k\tau,s} X_{\tau,t-s}, \qquad k = 1, 2, \ldots, p.$

Using arguments analogous to those for the bivariate case, we replace the periodogram ordinates by spectrum estimates and reduce the density of points in the summation to $2m$. The exponent of the likelihood function is then

(4.34) $\qquad \displaystyle\sum_{j=-m+1}^{m} \left\{ \sum_{k,l=1}^{p} g_{\eta\eta}^{kl}(\lambda_j) \left[\hat{g}_{y_l y_k}(\lambda_j) + \sum_{\tau,\nu=1}^{r_2} \sum_{s_1,s_2=-r_1}^{} a_{k\tau,s_1} a_{l\nu,s_2} \right. \right.$

$$\left. \left. \times \hat{g}_{x_\nu x_\tau}(\lambda_j) e^{i\lambda_j(s_2-s_1)} - 2 \sum_{\nu=1}^{N} \sum_{s=-r_1}^{r_2} a_{l\nu,s} \hat{g}_{x_\nu y_k}(\lambda_j) e^{i\lambda_j s} \right] \right\}.$$

Here, the subscripts x_τ, x_ν, y_k, and y_l refer to the sequences $\{X_{\tau,t}\}$, $\{X_{\nu,t}\}$, $\{Y_{k,t}\}$, and $\{Y_{l,t}\}$, respectively.

Maximizing (4.34) with respect to $a_{k\tau,s_1}$ leads to $Np(r_1 + r_2 + 1)$ equations of the form

(4.35) $\qquad \displaystyle\sum_{j=-m+1}^{m} \left\{ \sum_{l=1}^{p} g_{\eta\eta}^{kl}(\lambda_j) \left[\sum_{\nu=1}^{p} \sum_{s=-r_1}^{r_2} \tilde{a}_{l\nu,s} \hat{g}_{x_\nu x_\tau}(\lambda_j) e^{i\lambda_j(s-s_1)} \right. \right.$

$$\left. \left. - \hat{g}_{x_\tau y_l}(\lambda_j) e^{i\lambda_j s_1} \right] \right\} = 0,$$

$$k = 1, 2, \ldots, p; \quad \tau = 1, 2, \ldots, N;$$

$$s_1 = -r_1, -r_1 + 1, \ldots, 0, \ldots, r_2.$$

To solve for $\tilde{\mathbf{a}}$, a rearrangement of terms is necessary. Let $\mathbf{b}_s$ be an $Np \times 1$ vector with $a_{l\nu,s}$ in row $N(l-1)+\nu$, and let $h_{xy}(\lambda)$ be an $Np \times 1$ vector with $g_{x_\tau y_l}(\lambda)$ in row $N(l-1)+\tau$. Then we may write (4.35) as

(4.36) $\qquad \displaystyle (2m)^{-1} \sum_{j=-m+1}^{m} \left\{ [\mathbf{g}_{\eta\eta}^{-1}(\lambda_j) \otimes \hat{\mathbf{g}}_{xx}(-\lambda_j)] \sum_{s=-r_1}^{r_2} \tilde{\mathbf{b}}_s e^{i\lambda_j(s-s_1)} \right\}$

$$= (2m)^{-1} \sum_{j=-m+1}^{m} \left\{ [\mathbf{g}_{\eta\eta}^{-1}(\lambda_j) \otimes \mathbf{I}_N] \hat{\mathbf{h}}_{xy}(\lambda_j) e^{i\lambda_j s_1} \right\},$$

$$s_1 = -r_1, -r_1 + 1, \ldots, 0, \ldots, r_2,$$

where $\mathbf{I}_N$ denotes an $N \times N$ identity matrix.

We define an $Np(r_1 + r_2 + 1) \times 1$ vector

$$(4.37) \qquad \mathbf{b} = \begin{bmatrix} \mathbf{b}_{-r_1} \\ \mathbf{b}_{-r_1+1} \\ \vdots \\ \mathbf{b}_0 \\ \vdots \\ \mathbf{b}_{r_2} \end{bmatrix},$$

an $(r_1 + r_2 + 1) \times (r_1 + r_2 + 1)$ matrix $\mathbf{W}(\lambda_j)$ with $e^{i\lambda_j(s - s_1)}$ in row $(s + r_1 + 1)$ and column $(s + r_1 + 1)$, and an $(r_1 + r_2 + 1) \times 1$ vector $\mathbf{Z}(\lambda_j)$ with $e^{i\lambda_j s_1}$ in row $(s_1 + r_1 + 1)$. Expressions (4.35) and (4.36) may now be expressed as

$$(4.38a) \qquad \left\{ (2m)^{-1} \sum_{j=-m+1}^{m} \mathbf{W}(\lambda_j) \otimes [\mathbf{g}_{\eta\eta}^{-1}(\lambda_j) \otimes \hat{\mathbf{g}}_{xx}(-\lambda_j)] \right\} \tilde{\mathbf{b}}$$

$$= (2m)^{-1} \sum_{j=-m+1}^{m} \{ \mathbf{Z}(\lambda_j) \otimes [\mathbf{g}_{\eta\eta}^{-1}(\lambda_j) \otimes \mathbf{I}_N] \hat{\mathbf{h}}_{xy}(\lambda_j) \}$$

so that our estimator of $\mathbf{b}$ is

$$(4.38b) \qquad \tilde{\mathbf{b}} = \left\{ (2m)^{-1} \sum_{j=-m+1}^{m} \mathbf{W}(\lambda_j) \otimes [\hat{\mathbf{g}}_{\eta\eta}^{-1}(\lambda_j) \otimes \hat{\mathbf{g}}_{xx}(-\lambda_j)] \right\}^{-1}$$

$$\times \left\{ (2m)^{-1} \sum_{j=-m+1}^{m} \mathbf{Z}(\lambda_j) \otimes [\hat{\mathbf{g}}_{\eta\eta}^{-1}(\lambda_j) \otimes \mathbf{I}_N] \hat{\mathbf{h}}_{xy}(\lambda_j) \right\},$$

$$(4.38c) \qquad \hat{\mathbf{g}}_{\eta\eta}(\lambda) = \hat{\mathbf{g}}_{yy}(\lambda) - \hat{\mathbf{g}}_{xy}^*(\lambda) \hat{\mathbf{g}}_{xx}^{-1}(\lambda) \hat{\mathbf{g}}_{xy}(\lambda).$$

Here, $\tilde{\mathbf{b}}$ has the estimate $\tilde{\mathbf{a}}_{k\tau, s_1}$ in row $N[p(s_1 + r_1) + k - 1] + \tau$.

Using limiting arguments similar to those for the bivariate case and applying the result in (3.40c), one may show that the $Np(r_1 + r_2 + 1)$ vector of estimates $T^{\frac{1}{2}}(\tilde{\mathbf{b}} - \mathbf{b})$ asymptotically has the normal distribution with mean zero and covariance matrix†

$$(4.39a) \qquad \left\{ (2\pi)^{-1} \int_{-\pi}^{\pi} \mathbf{W}(\lambda) \otimes [\mathbf{g}_{\eta\eta}^{-1}(\lambda) \otimes \mathbf{g}_{xx}(-\lambda)] \, d\lambda \right\}^{-1},$$

---

† See Hannan [38, p. 24].

which is the covariance matrix of the best linear unbiased estimator [32, p. 269]. As an estimator of (4.39a) we may use

$$(4.39b) \qquad \left\{ (2m)^{-1} \sum_{j=-m+1}^{m} \mathbf{W}(\lambda_j) \otimes [\hat{\mathbf{g}}_{\eta\eta}^{-1}(\lambda_j) \otimes \hat{\mathbf{g}}_{xx}(-\lambda)] \right\}^{-1}.$$

So far we have referred to $\{\mathbf{X}_t\}$ as a set of N exogenous sequences. More generally they are predetermined variables that, under certain conditions, may contain lagged endogenous variables. If $\{\boldsymbol{\eta}_t\}$ is a set of white noise processes, the above results continue to hold. If, however, a sequence η_l is autocorrelated, and lagged endogenous variables are included among the predetermined variables, then if $Y_{k,t-\nu}$ is correlated with η_l, the basic assumption of independence between the predetermined variables and the residuals is violated.

The estimator $\tilde{\mathbf{b}}$ in (4.38b) is identical with the generalized least-squares estimator, with a consistent estimator replacing $\mathbf{g}_{\eta\eta}(\lambda)$. The estimator is therefore appealing for nonnormal processes as well.

4.13. Single Equation Multiple Regression

It is naturally desirable to reduce the dimensionality of the estimation problem whenever possible. Suppose that

$$(4.40) \qquad g_{\eta_k\eta_l}(\lambda) = 0 \qquad k \neq l.$$

Then we may estimate the coefficients of each of the p equations in (2.49) separately, still retaining the desirable sampling properties. For the kth equation we have the $N(r_1 + r_2 + 1)$ estimation equations

$$(4.41a) \qquad \sum_{s=-r_1}^{r_2} \sum_{\nu=1}^{p} \tilde{a}_{k\nu,s} \sum_{j=-m+1}^{m} \hat{g}_{\eta_k\eta_k}^{-1}(\lambda_j) \hat{g}_{x_\nu x_\tau}(\lambda_j) e^{i\lambda_j(s-s_1)}$$

$$= \sum_{j=-m+1}^{m} \hat{g}_{\eta_k\eta_k}^{-1}(\lambda_j) \hat{g}_{x_\tau y_k}(\lambda_j) e^{i\lambda_j s_1},$$

$$\tau = 1, 2, \ldots, p; \quad s_1 = -r_1, -r_1 + 1, \ldots, 0, \ldots, r_2.$$

We define an $N(r_1 + r_2 + 1) \times 1$ vector $\mathbf{b}_k$ with $a_{k\nu,s}$ in row $p(s + r_1) + \nu$ and an $N(r_1 + r_2 + 1) \times 1$ vector $\mathbf{g}_{x\nu_k}(\lambda)$ with $g_{x_\tau y_k}(\lambda)$ in row τ. Then $\tilde{\mathbf{b}}_k$ may be written as

$$(4.41b) \qquad \tilde{\mathbf{b}}_k = \left\{ \sum_{j=-m+1}^{m} \hat{g}_{\eta_k\eta_k}^{-1}(\lambda_j)[\mathbf{W}(\lambda_j) \otimes \hat{\mathbf{g}}_{xx}(-\lambda_j)] \right\}^{-1}$$

$$\times \left\{ \sum_{j=-m+1}^{m} \hat{g}_{\eta_k\eta_k}^{-1}(\lambda_j)[Z(\lambda_j) \otimes \hat{\mathbf{g}}_{x\nu_k}(\lambda_j)] \right\}$$

with asymptotic covariance matrix

$$(4.42) \qquad \left\{ (2\pi)^{-1}T \int_{-\pi}^{\pi} g_{\eta_k \eta_k}^{-1}(\lambda)[W(\lambda) \otimes g_{xx}(-\lambda)] \, d\lambda \right\}^{-1}.$$

If in addition $\{\eta_t\}$ are white noise processes, then (4.42) is

$$(4.43) \qquad \frac{R_{\eta_k \eta_k, 0}}{T} \begin{bmatrix} \mathbf{R}_{xx,0} & \mathbf{R}_{xx,-1} & \cdots & \mathbf{R}_{xx,-r_1-r_2-1} \\ \mathbf{R}_{xx,1} & \mathbf{R}_{xx,0} & & \vdots \\ \vdots & & \ddots & \\ \mathbf{R}_{xx,r_1+r_2+1} & & \cdots & \mathbf{R}_{xx,0} \end{bmatrix}^{-1},$$

which is the variance of the linear least-squares estimator.

4.14. Correlated White Noise Residual Processes

Suppose $\{\eta_t\}$ is a set of correlated white noise processes so that

$$(4.44) \qquad 2\pi g_{\eta\eta}(\lambda) = \mathbf{R}_{\eta\eta,0} \qquad 0 \leq |\lambda| \leq \pi,$$

a case often postulated in econometrics. As an estimate of $\mathbf{R}_{\eta\eta,0}$ we may use

$$(4.45) \qquad \hat{\mathbf{R}}_{\eta\eta,0} = \frac{\pi}{m}\left[\hat{g}_{\eta\eta}(0) + \hat{g}_{\eta\eta}(\pi) + 2\sum_{j=1}^{m-1} \hat{g}_{\eta\eta}(\lambda_j) \right],$$

which can be substituted for $\{g_{\eta\eta}(\lambda_j)\}$ in (4.38a).

It is not apparent that the spectral approach offers an advantage for estimating coefficients over more conventional econometric methods when (4.44) holds. The spectral approach, however, contributes prior to estimating coefficients. As described in Section 3.22, it enables the investigator to test the hypothesis that (4.44) is true, a freedom absent in time-domain analysis.

4.15. Constant Multivariate Signal-to-Noise Ratio

The extension of the constant signal-to-noise ratio assumption requires that

$$(4.46) \qquad \theta = [g_{\eta\eta}^{-1}(\lambda) \otimes g_{xx}(-\lambda)] \qquad 0 \leq |\lambda| \leq \pi,$$

where θ is an $Np \times Np$ matrix of constants. The asymptotically efficient estimator of $\tilde{\mathbf{b}}$ is then

$$(4.47) \quad \tilde{\mathbf{b}}_s = (2m)^{-1}\boldsymbol{\theta}^{-1} \sum_{j=-m+1}^{m} [\mathbf{g}_{\eta\eta}^{-1}(\lambda_j) \otimes \mathbf{I}_N]\hat{\mathbf{h}}_{xy}(\lambda_j)e^{i\lambda_j s}$$

$$= (2m)^{-1} \sum_{j=-m+1}^{m} [\mathbf{I}_p \otimes \mathbf{g}_{xx}^{-1}(-\lambda_j)]\hat{\mathbf{h}}_{xy}(\lambda_j)e^{i\lambda_j s},$$

$$s = -r_1, -r_1 + 1, \ldots, 0, \ldots, r_2,$$

since

$$(4.48) \quad \boldsymbol{\theta}^{-1}[\mathbf{g}_{\eta\eta}^{-1}(\lambda) \otimes \mathbf{I}_N] = [\mathbf{g}_{\eta\eta}(\lambda) \otimes \mathbf{g}_{xx}^{-1}(-\lambda)][\mathbf{g}_{\eta\eta}^{-1}(\lambda) \otimes \mathbf{I}_N]$$

$$= [\mathbf{I}_p \otimes \mathbf{g}_{xx}^{-1}(-\lambda)].$$

Expression (4.47) enables us to estimate additional lagged vector co-efficients without recomputing earlier ones. The asymptotic covariance matrix of the $r_1 + r_2 + 1$ vector coefficient estimators is

$$(4.49a) \quad \left\{2\pi T^{-1} \int_{-\pi}^{\pi} \mathbf{W}(\lambda) \otimes [\mathbf{g}_{\eta\eta}(\lambda) \otimes \mathbf{g}_{xx}^{-1}(-\lambda)] \, d\lambda\right\},$$

which equals $\boldsymbol{\theta}/T$ when $s_1 = s_2$ and vanishes elsewhere, provided that $\boldsymbol{\theta}$ is constant. As an estimate of (4.49a), we use

$$(4.49b) \quad (2mT)^{-1} \sum_{j=-m+1}^{m} \mathbf{W}(\lambda_j) \otimes [\hat{\mathbf{g}}_{\eta\eta}(\lambda_j) \otimes \hat{\mathbf{g}}_{xx}^{-1}(\lambda_j)].$$

Since it is doubtful that $\boldsymbol{\theta}$ is constant over $(-\pi, \pi)$, the estimator $\tilde{\mathbf{b}}_s$ is generally not efficient. Nevertheless, it has the advantage of not requiring the inversion of an $Np(r_1 + r_2 + 1) \times Np(r_1 + r_2 + 1)$ matrix as in (4.38b). To check the efficiency of (4.47), we may compare the determinant of the sample covariance matrix of (4.49a) with the determinant of (4.39a), but calculating the determinants consumes about as much computing time as calculating the inverses, the operation we hope to avoid.

While (4.46) may not hold, the estimator (4.47) can still assist us in making a judicious choice of r_1 and r_2, which are seldom known a priori. Consider the quantity†

$$(4.50a)$$

$$\sum_{s=-m+1}^{m} \tilde{\mathbf{b}}_s'\tilde{\mathbf{b}}_s = \sum_{s=-m+1}^{m} \sum_{l=1}^{p} \sum_{\tau=1}^{N} \tilde{a}_{l\tau,s}^2$$

† See Hannan [43, p. 417].

$$= (2m)^{-2} \sum_{s=-m+1}^{m} \sum_{j,k=-m+1}^{m} \{[\mathbf{I}_p \otimes \hat{\mathbf{g}}_{xx}^{-1}(-\lambda_j)]\hat{\mathbf{h}}_{xy}(\lambda_j)\}'$$

$$\times \{[\mathbf{I}_p \otimes \hat{\mathbf{g}}_{xx}^{-1}(-\lambda_k)]\hat{\mathbf{h}}_{xy}(\lambda_k)\} e^{i(\lambda_j+\lambda_k)s},$$

(4.50b)
$$\sum_{s=-m+1}^{m} \tilde{\mathbf{b}}_s'\tilde{\mathbf{b}}_s = (2m)^{-1} \sum_{j=-m+1}^{m} \hat{\mathbf{h}}_{xy}'(\lambda_j)$$

$$\times [\mathbf{I}_p \otimes \hat{\mathbf{g}}_{xx}^{-1}(\lambda_j)\hat{\mathbf{g}}_{xx}^{-1}(\lambda_j)]\hat{\mathbf{h}}_{xy}(-\lambda_j).$$

The term on the right in (4.50b) can be computed in advance. Then, using (4.47), one may estimate successive lagged vector coefficients, add the sum of squares of the new sample coefficients to the accumulated sum of squares of earlier ones, and compare this total with the right side of (4.50b). If the difference is small, then one may use the resulting negative and positive lag numbers as estimates of r_1 and r_2 and then estimate all the coefficients simultaneously from (4.38b). In this way one need invert only one large matrix, rather than several, in search of appropriate r_1 and r_2. The approach suggested here blends the computing advantage of (4.47) with the desirable sampling properties of (4.38b).

4.16. Consistency of a Multivariate Model

While the above approach has appeal in practice, it cannot relieve us of choosing r_1 and r_2 in an objective way consistent with statistical theory. Appropriate testing procedures remain to be developed; in the meantime, we can describe a rough test of the consistency of the model with the data. The suggested approach is to estimate r_1 and r_2 as above, estimate the coefficients using (4.38b), and use the following test to determine adequacy.

If the multivariate model is correctly specified, then $\{\mathbf{X}_t\}$ and $\{\boldsymbol{\eta}_t\}$ are mutually independent in the normal case. Consider the set of $(N + p)$ sequences

(4.51)
$$\mathbf{Z}_t = \begin{bmatrix} X_{1,t} \\ X_{2,t} \\ \vdots \\ X_{N,t} \\ \eta_{1,t} \\ \eta_{2,t} \\ \vdots \\ \eta_{p,t} \end{bmatrix}.$$

If $\{\mathbf{X}_t\}$ and $\{\boldsymbol{\eta}_t\}$ are mutually independent, then $\{Z_t\}$ has a spectrum matrix

(4.52)
$$\mathbf{g}_{zz}(\lambda) = \begin{bmatrix} \mathbf{g}_{xx}(\lambda) & 0 \\ 0 & \mathbf{g}_{\eta\eta}(\lambda) \end{bmatrix},$$

for all λ in $(0, \pi)$.

We again resort to a likelihood ratio test [4, Chap. 9], where the test statistic is

(4.53a) $\qquad V(\lambda) = (N + p)|\hat{\mathbf{g}}_{zz}(\lambda)|/[Np|\hat{\mathbf{g}}_{xx}(\lambda)| \, |\hat{\mathbf{g}}_{\eta\eta}(\lambda)|],$

with probability distribution such that

(4.53b) $\quad \Pr[-m_\lambda \log V(\lambda) \leq v] \sim \Pr(\chi_f^2 \leq v),$

(4.53c) $\qquad\qquad\qquad\qquad m_\lambda = 2n_\lambda - (N + p + 3)/2,$

(4.53d) $\qquad\qquad\qquad\qquad f = Np.$

Given $L + 1$ independent estimates on $(0, \pi)$, we may regard the test statistic

(4.54) $\quad [-4Ln_0 + (N + p + 3)(L + 1)/2] \sum_{j=0}^{L} \log V(\lambda_j)$

$$\lambda_j = \pi j/L$$

as being chi-square distributed with $Np(L + 1)$ degrees of freedom. To compute $\hat{\mathbf{g}}_{zz}(\lambda)$ we use the observations on $\{\mathbf{X}_t\}$ and the residuals formed by

(4.55) $\qquad\qquad\qquad \hat{\boldsymbol{\eta}}_t = \mathbf{Y}_t - \sum_{s=-r_1}^{r_2} \tilde{\mathbf{a}}_s \mathbf{X}_{t-s}.$

The quantity n_λ in (4.53c) is based on a sample size of $T - r_1 - r_2$ and on ignoring the substitution of $\tilde{\mathbf{a}}$ for $\mathbf{a}$.

The reader is again reminded of the approximating nature of sampling distributions in spectrum analysis and is cautioned to consider the above test as rough guidance. As before, it is suggested that L be set equal to $M/2$ where M is the number of lags and the spacing between successive estimates is π/M.

5 The Income-Consumption
Relationship

5.1. Introduction

In this chapter we use spectral methods to study the relationship between two comparatively short economic time series. The chapter serves two purposes. One is to offer a procedural example that will assist economists interested in using these methods; the other is to combat the view that spectral methods do not apply when the number of observations T is small. It is the ratio M/T, not T alone, that controls the reliability of our results. By appropriately prewhitening the time series, we can, in many cases, resolve the spectrum with a small enough M to make the ratio M/T acceptable from the reliability viewpoint.

For our time series we use quarterly United States disposable personal income and personal consumption expenditures in 1958 dollars. Both are seasonally adjusted. The sample period covers 76 quarters from 1947 through 1965.† The sample record length is then $T = 76$. We denote the income and consumption time series by X and Y respectively. Figure 28 shows these time series, both of which exhibit trend.

5.2. Spectrum Estimation

Our first concern was to prewhiten the time series in order to conserve on lags (M) when estimating spectra. Using the quasi-

---

† United States Department of Commerce [92].

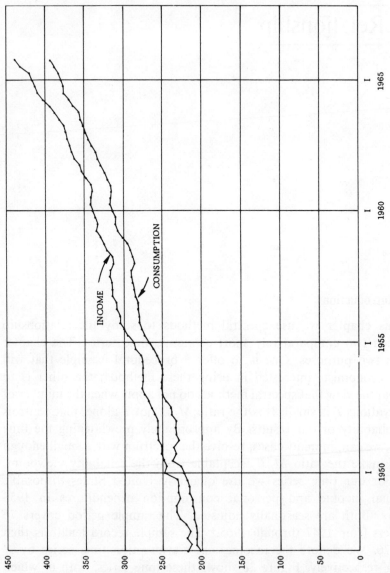

Fig. 28 United States quarterly income and consumption in 1958 dollars, seasonally adjusted, 1947–1965 (in billions of dollars)

differencing procedure described in Section 3.11, we denote the pre-whitened series by

(5.1a) $$X'_t = \sum_{s=0}^{n} \binom{n}{s}(-\alpha)^s X_{t-s},$$

$$(5.1\text{b}) \qquad Y'_t = \sum_{s=0}^{n} \binom{n}{s}(-\alpha)^s Y_{t-s},$$

$$t = n + 1, n + 2, \ldots, T.$$

The same prewhitening parameters, n and α, were applied to both series in order to take advantage of the invariance property of the distributed lag coefficients as discussed in Point 8 of Section 4.1.

The formulae for the sample covariance functions are

$$(5.2\text{a}) \qquad C_{x',\tau} = (T - n)^{-1} \sum_{t=n+1}^{T-\tau} (X'_t - \overline{X}')(X'_{t+\tau} - \overline{X}'),$$

$$(5.2\text{b}) \qquad C_{y',\tau} = (T - n)^{-1} \sum_{t=n+1}^{T-\tau} (Y'_t - \overline{Y}')(Y'_{t+\tau} - \overline{Y}'),$$

$$(5.2\text{c}) \qquad C_{x'y',\tau} = (T - n)^{-1} \sum_{t=n+1}^{T-\tau} (X'_t - \overline{X}')(Y'_{t+\tau} - \overline{Y}'),$$

$$(5.2\text{d}) \qquad C_{x'y',-\tau} = (T - n)^{-1} \sum_{t=n+1}^{T+\tau} (X'_{t+\tau} - \overline{X}')(Y'_t - \overline{Y}'),$$

$$(5.2\text{e}) \qquad \overline{X}' = (T - n)^{-1} \sum_{t=n+1}^{T} X'_t,$$

$$(5.2\text{f}) \qquad \overline{Y}' = (T - n)^{-1} \sum_{t=n+1}^{T} Y'_t,$$

$$\tau = 0, 1, \ldots, 40.$$

Second quasi-differencing ($n = 2$) was used. Earlier work had indicated that first quasi-differencing ($n = 1$) was generally inadequate for economic time series and that higher-order ($n > 2$) quasi-differencing was unnecessary if an appropriate α were chosen for flattening the spectrum. Rough prewhitened sample spectra were computed for $\alpha = 0.8, 0.85, 0.90,$ and 0.92. The formulae are

$$(5.3\text{a}) \qquad \hat{h}_{x'M}(\lambda_j) = (2\pi)^{-1}\left(C_{x',0} + 2\sum_{\tau=1}^{M} k_{M,\tau}C_{x',\tau} \cos \lambda_j\tau\right),$$

$$(5.3\text{b}) \qquad \hat{h}_{y'M}(\lambda_j) = (2\pi)^{-1}\left(C_{y',0} + 2\sum_{\tau=1}^{M} k_{M,\tau}C_{y',\tau} \cos \lambda_j\tau\right),$$

$$\lambda_j = \pi j/20, \quad j = 0, 1, \ldots, 20; \quad M = 10, 20, 40.$$

The Parzen window was used throughout the analysis.

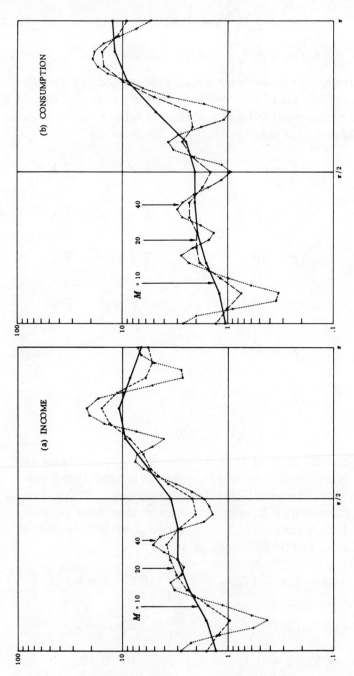

Fig. 29 Prewhitened sample spectra, $T = 74$, $\alpha = .92$

These estimates were adjusted for sample mean bias so that the prewhitened spectra became†

$$(5.4a) \qquad \hat{g}_{x'M}(0) = \hat{h}_{x'M}(0)/[1 - 2\pi K_M(0)/T],$$

$$(5.4b) \qquad \hat{g}_{y'M}(0) = \hat{h}_{y'M}(0)/[1 - 2\pi K_M(0)/T],$$

$$(5.4c) \qquad \hat{g}_{x'M}(\lambda_j) = \hat{h}_{x'M}(\lambda_j) + 2\pi K_M(\lambda_j)\hat{g}_{x'M}(0)/T,$$

$$(5.4d) \qquad \hat{g}_{y'M}(\lambda_j) = \hat{h}_{y'M}(\lambda_j) + 2\pi K_M(\lambda_j)\hat{g}_{y'M}(0)/T,$$

$$(5.5) \qquad 2\pi K_M(\lambda) = \sum_{\tau=-M}^{M} k_{M,\tau} \cos \lambda\tau = 1 + 2 \sum_{\tau=1}^{M} k_{M,\tau} \cos \lambda\tau.$$

The estimates were plotted and examined to choose appropriate α and M. Figure 29 shows the prewhitened sample spectra for $\alpha = 0.92$ and $M = 10, 20, 40$. This choice of α appears to remove the low-frequency content of the two spectra adequately, so we decided to use it in the remaining analysis. Notice that the sample spectra for 10 lags are rather smooth compared to those for 20 and 40 lags. As a compromise between resolution and stability we chose 20 lags. This choice gave us a bandwidth of

$$(5.6a) \qquad \beta_M = 8\pi/(3M) = 2\pi/15$$

and a variance of

$$(5.6b) \qquad \text{var } [\hat{g}(\lambda)] \sim .542[\delta_\lambda M/(T - 2)]g^2(\lambda)$$

$$= .146\delta_\lambda g^2(\lambda),$$

assuming X' and Y' are normal processes, and dropping the subscript M.

The recolored sample spectra were derived using the formulae

$$(5.7a) \qquad \hat{g}_x(\lambda) = \hat{g}_{x'}(\lambda)H(\lambda),$$

$$(5.7b) \qquad \hat{g}_y(\lambda) = \hat{g}_{y'}(\lambda)H(\lambda),$$

$$(5.7c) \qquad H(\lambda) = 1/(1 - 2\alpha \cos \lambda + \alpha^2)^n.$$

Figure 30 shows these spectra for income and consumption. Since the sampling interval is the quarter, the seasonal frequencies are $\pi/2$ and π, corresponding to the yearly and six-month seasonal cycles, respectively. The absence of peaks at these frequencies is to be expected since both time series are seasonally adjusted.

---

† See Section 3.5.

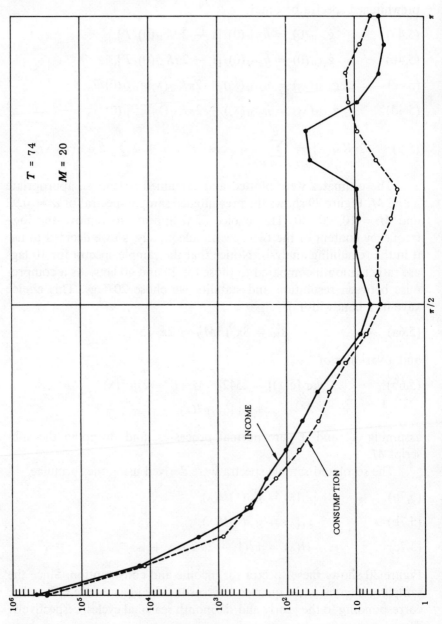

Fig. 30 Sample spectra

Once α and M had been chosen, we computed rough prewhitened sample co- and quadrature-spectra using the formulae

(5.8a) $\qquad \hat{h}_{c'}(\lambda_j) = (2\pi)^{-1} \sum_{\tau=-M}^{M} k_{M,\tau} C_{x'y',\tau+\nu} \cos \lambda_j \tau,$

(5.8b) $\qquad \hat{h}_{q'}(\lambda_j) = (2\pi)^{-1} \sum_{\tau=-M}^{M} k_{M,\tau} C_{x'y',\tau+\nu} \sin \lambda_j \tau,$

$$M = 20.$$

Here we set $\nu = 0$, since an examination of the sample covariance function $C_{x'y'}$ revealed no unusually large peak for $\nu \neq 0$ that would distort our estimates as described in Section 3.14. We adjusted these estimates to correct for bias so that our prewhitened sample cross spectrum is

(5.9a) $\qquad \hat{c}'(0) = \hat{h}_{c'}(0)/[1 - 2\pi K_M(0)/T],$

(5.9b) $\qquad \hat{c}'(\lambda_j) = \hat{h}_{c'}(\lambda_j) + 2\pi K_M(\lambda_j)\hat{c}'(0)/T,$

(5.9c) $\qquad \hat{q}'(0) = \hat{q}'(\pi) = 0,$

(5.9d) $\qquad \hat{q}'(\lambda_j) = \hat{h}_{q'}(\lambda_j)/w_M(\lambda_j),$

(5.9e) $\qquad w_M(\lambda) = (4/\pi) \sum_{\tau=1}^{M} [k_{M,\tau} \sin \lambda\tau \sin^2 (\pi\tau/2)/\tau],$

$$j = 1, 2, \ldots, M - 1.$$

Had ν been nonzero, a further adjustment would have been required so that the co- and quadrature-spectrum estimates would have been

(5.10a) $\qquad \hat{c}'(\lambda) \cos \lambda\nu - \hat{q}'(\lambda) \sin \lambda\nu,$

(5.10b) $\qquad \hat{q}'(\lambda) \cos \lambda\nu + \hat{c}'(\lambda) \sin \lambda\nu,$

respectively.

The sample coherence, gain, and phase angle were computed from

(5.11a) $\qquad \hat{\gamma}(\lambda) = \{[\hat{c}'^2(\lambda) + \hat{q}'^2(\lambda)]/[\hat{g}_{x'}(\lambda)\hat{g}_{y'}(\lambda)]\}^{\frac{1}{2}},$

(5.11b) $\qquad \hat{G}(\lambda) = [\hat{c}'^2(\lambda) + \hat{q}'^2(\lambda)]^{\frac{1}{2}}/\hat{g}_{x'}(\lambda),$

(5.11c) $\qquad \hat{\phi}(\lambda) = \tan^{-1} [-\hat{q}'(\lambda)/\hat{c}'(\lambda)],$

respectively, and are shown in Fig. 31. Notice the dips in both the coherence and gain functions at the seasonal frequencies. No attempt was made here to correct for the bias described in Section 3.17.

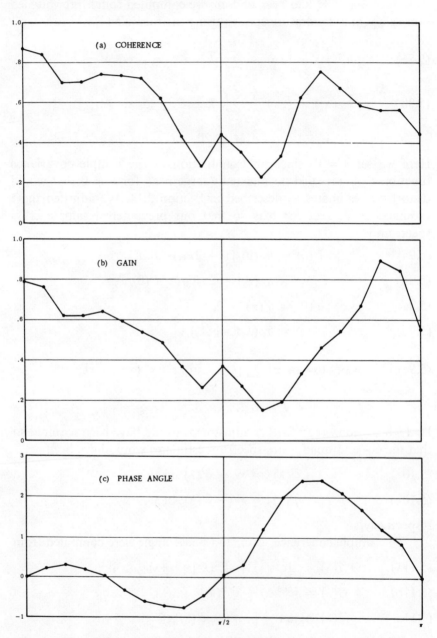

Fig. 31 Sample coherence, gain, and phase angle of income and consumption

5.3. The Adaptive Model

In econometrics the adaptive model†

$$(5.12) \qquad Y_t = \theta + \alpha \sum_{s=0}^{\infty} \beta^s X_{t-s} + \eta_t \qquad |\beta| < 1$$

has often been used to study the income-consumption relationship. Here, X and η are mutually uncorrelated sequences and θ, α, and β are constants to be determined. Since α is the one period change in Y due to a unit change in X, we may regard α as the short-term marginal propensity to consume (MPC). Suppose that the income mean changes by one unit. Then the overall increase in consumption is $\alpha/(1 - \beta)$, which is the long-term MPC.

The corresponding frequency response function is

$$(5.13a) \qquad A(\lambda) = \alpha/(1 - \beta e^{-i\lambda}),$$

so that the gain and phase angle are

$$(5.13b) \qquad G(\lambda) = \alpha/(1 - 2\beta \cos \lambda + \beta^2),$$

$$(5.13c) \qquad \phi(\lambda) = \tan^{-1}[-\sin \lambda/(1 - \beta \cos \lambda)],$$

respectively.

Notice that the gain is a monotonically decreasing function of λ. The sample gain shown in Fig. 31b clearly has a strong peak in the high-frequency interval. On the basis of this visual inspection of the gain, it seems inappropriate to fit the adaptive model using quarterly data.

As an alternative to the adaptive model, we define

$$(5.14a) \qquad W_s = X_{t+1} + X_t,$$

$$(5.14b) \qquad Z_s = Y_{t+1} + Y_t \qquad t = 1, 3, 5, \ldots,$$

$$(5.14c) \qquad s = (t+1)/2, \ldots, \qquad S = T/2.$$

Here W_s and Z_s are semiannual income and consumption, respectively, and the indices s and t are in six-month intervals and quarters, respectively. We have the covariance functions

$$(5.15a) \qquad R_{w,\tau} = R_{x, 2\tau} + 2R_{x, 2\tau} + R_{x, 2\tau-1},$$

$$(5.15b) \qquad R_{z,\tau} = R_{y, 2\tau+1} + 2R_{y, 2\tau} + R_{y, 2\tau-1},$$

$$(5.15c) \qquad R_{wz,\tau} = R_{xy, 2\tau+1} + 2R_{xy, 2\tau} + R_{xy, 2\tau-1},$$

† See Section 4.9.

noting that

(5.16a) $$W_{s+\tau} = X_{2(s+\tau)} + X_{2(s+\tau)-1},$$

(5.16b) $$Z_{s+\tau} = Y_{2(s+\tau)} + Y_{2(s+\tau)-1}.$$

We may also write

(5.17a)

$$
\begin{aligned}
\int_{-\pi}^{\pi} g_w(\lambda) e^{i\lambda\tau} \, d\lambda &= \int_{-\pi}^{\pi} g_x(\lambda)(e^{i\lambda} + 2 + e^{-i\lambda}) e^{i2\lambda\tau} \, d\lambda \\
&= 4 \int_{-\pi}^{\pi} g_x(\lambda) \cos^2 (\lambda/2) e^{i2\lambda\tau} \, d\lambda \\
&= \int_{-2\pi}^{2\pi} g_x(\lambda/2) \cos^2 (\lambda/4) e^{i\lambda\tau} \, d\lambda \\
&= \sum_{j=-1}^{1} \int_{-\pi}^{\pi} g_x(\lambda/2 + j\pi/2) \cos^2 (\lambda/4) e^{i\lambda\tau} \, d\lambda,
\end{aligned}
$$

so that

(5.17b) $$g_w(\lambda) = \sum_{j=-1}^{1} g_x(\lambda/2 + j\pi/2) \cos^2 (\lambda/4),$$

and similarly,

(5.17c) $$g_z(\lambda) = \sum_{j=-1}^{1} g_y(\lambda/2 + j\pi/2) \cos^2 (\lambda/4),$$

(5.17d) $$g_{wz}(\lambda) = \sum_{j=-1}^{1} g_{xy}(\lambda/2 + j\pi/2) \cos^2 (\lambda/4).$$

We see that the spectra g_w, g_z, and g_{wz} contain aliases.[†] The aliasing involves components in $g_x(\omega)$, $g_y(\omega)$, $g_{xy}(\omega)$ for $|\omega| > \pi/2$. From Fig. 30 we note that $\hat{g}_x$ and $\hat{g}_y$ contribute relatively little to the spectra for $\omega > \pi/2$ as compared with the contributions for $\omega \leq \pi/2$. We may then write

(5.18a) $$g_w(\lambda) \sim g_x(\lambda/2) \cos^2 (\lambda/4),$$

(5.18b) $$g_z(\lambda) \sim g_y(\lambda/2) \cos^2 (\lambda/4),$$

(5.18c) $$g_{wz}(\lambda) \sim g_{xy}(\lambda/2) \cos^2 (\lambda/4), \qquad 0 \leq |\lambda| \leq \pi.$$

Our semiannual adaptive model is

(5.19) $$W_s = \delta + \alpha \sum_{s=0}^{\infty} \beta^s Z_{s-\nu} + \xi_s,$$

---

† See Section 2.13.

where Z and ξ are mutually uncorrelated sequences. To apply the Hannan estimation procedure (Section 4.9), we require the ratios

(5.20a) $$\hat{g}_w(\lambda)/\hat{g}_\xi(\lambda) \sim \hat{g}_x(\lambda/2)/\hat{g}_\eta(\lambda/2),$$

(5.20b) $$\hat{g}_{wz}(\lambda)/\hat{g}_\xi(\lambda) \sim \hat{g}_{xy}(\lambda/2)/\hat{g}_\eta(\lambda/2).$$

We see that no additional spectrum estimates are required.

Before proceeding with the estimation of α, β, and φ, it is instructive to compare the sample frequency response function with the one corresponding to the adaptive specification. We have

(5.21a) $$g_{xy}(\lambda) = g_x(\lambda)[\alpha/(1 - \beta e^{-i\lambda})].$$

Suppose β is small compared to unity. Then

(5.21b) $$g_{xy}(\lambda) \sim g_x(\lambda)\alpha(1 + \beta e^{-i\lambda}),$$

so that

(5.21c) $$c(\lambda)/g_x(\lambda) \sim \alpha(1 + \beta \cos \lambda),$$

(5.21d) $$q(\lambda)/g_x(\lambda) \sim \alpha\beta \sin \lambda.$$

Since

(5.21e) $$\sin \lambda = \cos (\lambda - \pi/2),$$

we may write

(5.21f) $$c(\lambda - \pi/2)/g_x(\lambda) \sim \alpha + q(\lambda)/g_x(\lambda).$$

The validity of this relationship may be checked using Fig. 32. The jth point corresponds to frequency $j\pi/10$, which is measured in radians per six months. Except for points 9 and 10, the points appear to approximate a straight line, with unit slope thus supporting the adaptive model with small β. The distortion at points 9 and 10 is probably due to β being larger than we had assumed.

5.4. Estimation of Coefficients

To estimate α and β, we used the ordinates $2\pi/M$ radians per quarter apart over $(0, \pi/2)$. Since $M = 20$, this gives us six ordinates over which to average. The frequency axis was relabeled so that $2\pi j/M$ radians per quarter was defined as $\pi j/M$ radians per six months.

The estimated coefficients are

(5.22a) $$\tilde{\alpha} = 0.537 \qquad \tilde{\beta} = 0.217,$$

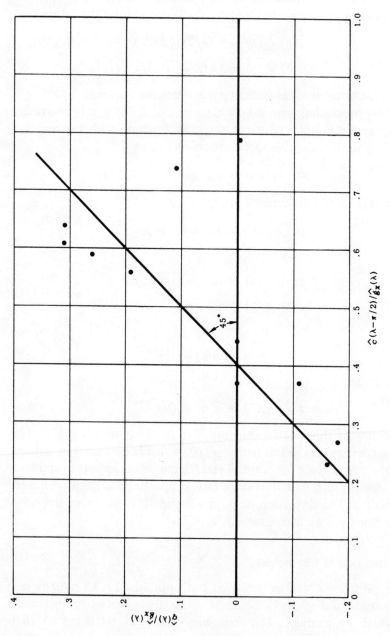

Fig. 32 $\hat{c}(\lambda - \pi/2)/\hat{g}_x(\lambda)$ versus $\hat{q}(\lambda)/\hat{g}_x(\lambda)$

with sample covariance matrix

$$(5.22b) \qquad \begin{bmatrix} .004965 & -.001776 \\ -.002638 & .013453 \end{bmatrix}.$$

The sample standard errors of $\tilde{\alpha}$ and $\tilde{\beta}$ are 0.070 and 0.116, respectively. The asymmetry of the sample covariance matrix is due to sampling errors since, in the limit, the matrix is symmetric (Section 4.9). We use the sample covariance -0.002638 since it theoretically has a smaller variance than -0.001776.

Before estimating the constant θ we note that

$$(5.23a) \qquad \tilde{\alpha}/(1 - \tilde{\beta}) \sim \alpha/(1 - \beta) + (\tilde{\alpha} - \alpha)/(1 - \beta)$$
$$- \alpha(\tilde{\beta} - \beta)/(1 - \beta)^2 - (\tilde{\alpha} - \alpha)(\tilde{\beta} - \beta)/(1 - \beta)^2$$
$$+ \alpha(\tilde{\beta} - \beta)^2/(1 - \beta)^3,$$

so that

$$(5.23b) \qquad E[\tilde{\alpha}/(1 - \tilde{\beta})] \sim \alpha/(1 - \beta) - \text{cov}\,(\tilde{\alpha}, \tilde{\beta})/(1 - \beta)^2$$
$$+ \alpha\,\text{var}\,(\tilde{\beta})/(1 - \beta)^3.$$

We estimated $\alpha/(1 - \beta)$ by

$$(5.23c) \qquad \tilde{\varphi} = [\tilde{\alpha} + \widehat{\text{cov}}\,(\tilde{\alpha}, \tilde{\beta})/(1 - \tilde{\beta})$$
$$- \tilde{\alpha}\,\widehat{\text{var}}\,(\tilde{\beta})/(1 - \tilde{\beta})^2]/(1 - \tilde{\beta}) = .666,$$

with variance

$$(5.23d) \qquad \text{var}\,(\tilde{\varphi}) = [\text{var}\,(\tilde{\alpha}) - 2\alpha\,\text{cov}\,(\tilde{\alpha}, \tilde{\beta})/(1 - \beta)$$
$$+ \alpha^2\,\text{var}\,(\tilde{\beta})/(1 - \beta)^2]/(1 - \beta)^2$$

estimated by

$$(5.23e) \qquad \widehat{\text{var}}\,(\tilde{\varphi}) = [\widehat{\text{var}}\,(\tilde{\alpha}) - 2\tilde{\alpha}\,\widehat{\text{cov}}\,(\tilde{\alpha}, \tilde{\beta})/(1 - \tilde{\beta})$$
$$+ \tilde{\alpha}^2\,\widehat{\text{var}}\,(\tilde{\beta})/(1 - \tilde{\beta})^2]/(1 - \tilde{\beta})^2 = .024321.$$

The constant θ was estimated by

$$(5.24a) \qquad \tilde{\theta} = \overline{Z} - \tilde{\varphi}\overline{W} = 38.9,$$

$$(5.24b) \qquad \overline{W} = S^{-1} \sum_{s=1}^{S} W_s = 153.9,$$

$$(5.24c) \qquad \overline{Z} = S^{-1} \sum_{s=1}^{S} Z_s = 141.377.$$

These data are given at semiannual rates.

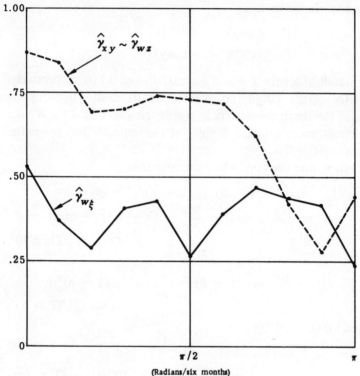

(Radians/six months)

Fig. 33 Sample coherence of income and consumption

5.5. Testing the Validity of the Model

The consumption residuals were computed from

$$(5.25) \qquad U_s = \xi_s - \tilde{\beta}\xi_{s-1} = Z_s - \tilde{\beta}Z_{s-1}\tilde{\alpha}W_s - \tilde{\theta}(1 - \tilde{\beta}),$$
$$s = 2, 3, \ldots, S.$$

Figure 33 shows the sample coherence between W and ξ. Second quasi-differencing with $\alpha = 0.85$ was used to prewhiten the data, and 10 lags were used in the estimation.† The reliability of the sample coherence is measured by

$$(5.26) \qquad n_\lambda = 1/(\delta_\lambda \Psi_{S-3,M}) = 6.5/\delta_\lambda \sim 6/\delta_\lambda,$$

assuming W and U are normal processes. Notice that $\hat{\gamma}_{w\xi}$ is much lower than $\hat{\gamma}_{xy}$ in the low-frequency range.

---

† The choice of $\alpha = 0.85$ was to compensate for the fact that $\xi_s - \tilde{\beta}\xi_{s-1}$ has a flatter spectrum than ξ.

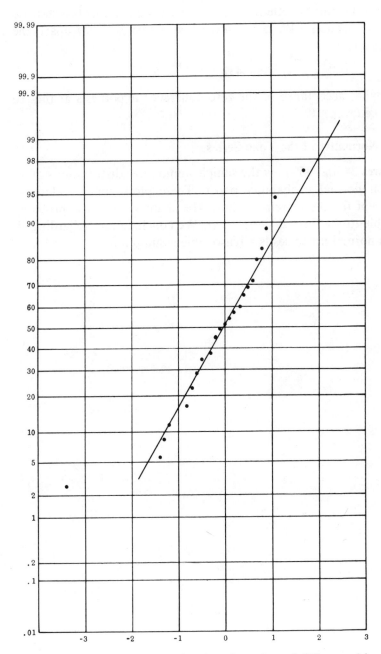

Fig. 34 Sample distribution function of quasi-differenced income

To test the null hypothesis, $\gamma_{w\xi} = 0$, we computed the test statistic described in Section 3.18. Using successive estimates $2\pi/M$ apart, the test statistic is

$$(5.27) \qquad\qquad \Delta_L \sim -.437, \qquad L = 5,$$

which implies acceptance of the zero coherence hypothesis at the .10 significance level.

5.6. The Normality of the Time Series

Figures 34 and 35 show the sample cumulative distributions of W and U on normal probability paper. These plots appear relatively linear except for the extreme points. The slight skewness is probably attributable to the limited sample size. We conclude that regarding W and U as normal processes is a reasonable assumption.

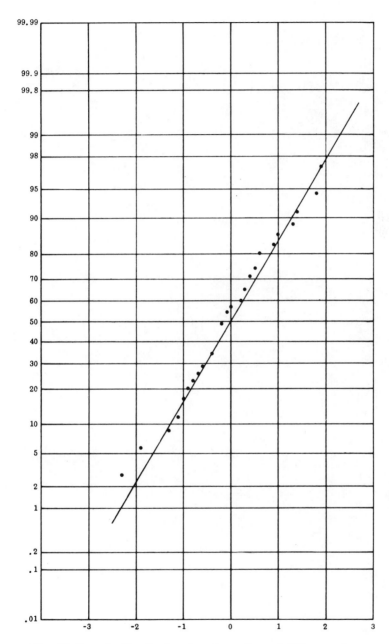

Fig. 35 Sample distribution function of quasi-differenced consumption

References

Index

References

1. Adelman, I., "Long Cycles—Fact or Artifact?" *Am. Econ. Rev.*, Vol. 55, No. 3, June 1965, pp. 444–463.

2. Amemiya, T., and W. Fuller, *A Comparative Study of Alternative Estimators in a Distributed-Lag Model*, Institute for Mathematical Studies in the Social Sciences, Stanford University, Technical Report 12, National Science Foundation Grant, GS-142, June 23, 1965.

3. Amos, D. E., and L. H. Koopmans, *Tables of the Distribution of the Coefficient of Coherence for Stationary Bivariate Gaussian Processes*, Sandia Corporation Monograph, SCR-483, March 1963.

4. Anderson, T. W., *An Introduction to Multivariate Statistical Analysis*, John Wiley and Sons, New York, 1958.

5. Bartlett, M. S., *An Introduction to Stochastic Processes*, Cambridge University Press, London, 1961.

6. ———, "Some Remarks on the Analysis of Time-Series," *Biometrika*, Vol. 54, Nos. 1 and 2, 1967, pp. 25–38.

7. Bendat, J. S., and A. G. Piersol, *Measurement and Analysis of Random Data*, John Wiley and Sons, New York, 1966.

8. Beveridge, W. H., "Wheat Prices and Rainfall in Western Europe," *J. Roy. Stat. Soc.*, Vol. 85, 1922, pp. 412–459.

9. ———, "Weather and Harvest Cycles," *Econ. J.*, Vol. 31, December 1921, pp. 429–452.

10. Bingham, C., M. D. Godfrey, and J. W. Tukey, "Modern Techniques of Power Spectrum Estimation," *IEEE Transactions on Audio and Electroacoustics*, Vol. AU-15, No. 2, June 1967, pp. 56–66.

11. Blackman, R. B., and J. W. Tukey, *The Measurement of Power Spectra*, Dover Publications, New York, 1958.

12. Box, G. E. P., and G. M. Jenkins, *Recent Advances in Forecasting and Control*, presented at the European Meeting of Statisticians, London, September 1966 (mimeograph).

13. Cooley, J. W., and J. W. Tukey, "An Algorithm for the Machine Calculation of Fourier Series," *Math. Comput.*, Vol. 19, 1965, pp. 297–301.

14. Couts, D., D. Grether, and M. Nerlove, "Forecasting Non-Stationary Economic Time Series," *Management Science*, Vol. 13, No. 1, September 1966, pp. 1–21.

15. Cox, D. R., and H. D. Miller, *The Theory of Stochastic Processes*, John Wiley and Sons, New York, 1965.

16. Cramér, H., *Mathematical Methods of Statistics*, Princeton University Press, Princeton, New Jersey, 1946.

17. Danielson, G. C., and C. Lanczos, "Some Improvements in Practical Fourier Analysis and Their Application to X-ray Scattering from Liquids," *J. Frank. Inst.*, Vol. 233, 1924, pp. 365–380, 435–452.

18. Davenport, W. B., Jr., and W. L. Root, *Random Signals and Noise*, McGraw-Hill Book Company, Inc., New York, 1958.

19. Davis, H. T., *The Analysis of Economic Time Series*, Principia Press, Bloomington, Indiana, 1941.

20. Doob, J. L., *Stochastic Processes*, John Wiley and Sons, New York, 1953.

21. Durbin, J., "Trend Elimination by Moving Average and Variate Difference Filters," *Bull. inst. intern. de stat.*, Vol. 39, 1961, pp. 130–141.

22. Enochson, L. D., *Frequency Response Functions and Coherence Functions for Multiple Input Linear Systems*, National Aeronautics and Space Administration, NASA CR-32, Washington, D.C., April 1964.

23. Enochson, L. D., and N. R. Goodman, *Gaussian Approximations to the Distribution of Sample Coherence*, Research and Technology Division, AFSC, AFFDL TR 65-57, Wright-Patterson AFB, Ohio, June 1965.

24. Friedman, M., *A Theory of the Consumption Function*, Princeton University Press, Princeton, New Jersey, 1957.

25. Goodman, N. R., *On the Joint Estimation of the Spectra, Cospectrum and Quadrature Spectrum of a Two-Dimensional Stationary Gaussian Process*, Scientific Paper 10, Engineering Statistical Laboratory, College of Engineering, New York University, March 1957 (mimeograph).

26. ———, "Statistical Analysis Based on a Certain Multivariate Complex Gaussian Distribution," *Ann. Math. Stat.*, Vol. 34, No. 1, March 1963, pp. 152–177.

27. ———, "The Distribution of the Determinant of a Complex Wishart Distributed Matrix," *Ann. Math. Stat.*, Vol. 34, No. 1, March 1963, pp. 178–180.

28. ———, *Measurement of Matrix Frequency Response Functions and Multiple Coherence Functions*, Research and Technology Division, AFSC, AFFDL TR 65-56, Wright-Patterson AFB, Ohio, June 1965.

29. Gordon Committee Report, President's Committee to Appraise Employment and Unemployment Statistics, R. A. Gordon, Chairman, *Measuring Employment and Unemployment*, United States Government Printing Office, Washington, D.C., 1962.

30. Granger, C. W. J., and M. Hatanaka, *Spectral Analysis of Economic Time Series*, Princeton University Press, Princeton, New Jersey, 1964.

31. Granger, C. W. J., and O. Morgenstern, "Spectral Analysis of New York Stock Market Prices," *Kyklos*, No. 16, 1963, pp. 1–27.

32. Grenander, U., and M. Rosenblatt, *Statistical Analysis of Stationary Time Series*, John Wiley and Sons, New York, 1957.

33. Grenander, U., and G. Szegö, *Toeplitz Forms and Their Applications*, University of California Press, Berkeley and Los Angeles, 1958.

34. Halmos, P. R., *Finite-Dimensional Vector Spaces*, Van Nostrand, Princeton, New Jersey, 1958.

35. Hamon, B. V., and E. J, Hannan, "Estimating Relations Between Time Series," *J. Geophysical Res.*, Vol. 68, No. 21, November 1963, pp. 6033–6041.

36. Hannan, E. J., *Time Series Analysis*, Methuen, London, 1960.

37. ———, "The Estimation of Seasonal Variation," *Australian J. Stat.*, Vol. 2, 1960, pp. 1–15.

38. ———, "Regression for Time Series," in *Proceedings of the Symposium on Time Series Analysis*, Brown University, June 11–14, 1962, M. Rosenblatt (ed.), John Wiley and Sons, New York, 1963.

39. ———, "The Estimation of Seasonal Variation in Economic Time Series," *J. Am. Stat. Assoc.*, Vol. 58, No. 301, March 1963, pp. 31–44.

40. ———, "The Estimation of a Changing Seasonal Pattern," *J. Am. Stat. Assoc.*, Vol. 59, No. 308, December 1964, pp. 1063–1077.

41. ———, *Notes on Time Series Analysis, Part II*, lectures by E. J. Hannan, 1964–1965, notes recorded by C. A. Rohde, Department of Statistics, The Johns Hopkins University, Baltimore, Copyright applied for.

42. ———, "The Estimation of Relationships Involving Distributed Lags," *Econometrica*, Vol. 33, No. 1, January 1965, pp. 206–224.

43. ———, "The Estimation of a Lagged Regression Relation," *Biometrika*, Vol. 54, Nos. 3 and 4, 1967, pp. 409–418.

44. Hatanaka, M., and E. P. Howrey, *Another View of the Long Swing: Comments on Adelman's Study of Long Cycles*, Econometric Research Program, Research Memorandum 77, Princeton University, July 1965.

45. Hood, W. C., and T. C. Koopmans (eds.), *Studies in Econometric Method*, John Wiley and Sons, New York, 1953.

46. Howrey, E. P., *A Spectral Analysis of the Long-Swing Hypothesis*, Econometric Research Program, Research Memorandum 78, Princeton University, August 1965.

47. Jenkins, G. M., "General Considerations in the Analysis of Spectra," *Technometrics*, Vol. 3, No. 2, May 1961, pp. 133–166.

48. ———, "Comments on the Discussions of Messrs. Tukey and Goodman," *Technometrics*, Vol. 3, No. 2, May 1961, pp. 229–232.

49. ———, "Cross-Spectral Analysis and the Estimation of Open Loop Transfer Functions," in *Proceedings of the Symposium on Time Series Analysis*, Brown University, June 11–14, 1962, M. Rosenblatt (ed.), John Wiley and Sons, New York, 1963.

50. ———, "Some Examples of and Comments on Spectral Analysis," in *Proceedings of the IBM Scientific Computing Symposium on Statistics*, IBM Data Processing Division, White Plains, New York, 1965.

51. Jevons, W. S., *Investigations in Currency and Finance*, Macmillan and Company, London, 1884.

52. Kendall, M. G., *Contributions to the Study of Oscillatory Time Series*, Occasional Paper IX, National Institute of Economic and Social Research, Cambridge University Press, Cambridge, 1946.

53. ———, "Note on Bias in the Estimation of Autocorrelation," *Biometrika*, Vol. 41, Parts 3 and 4, 1954, pp. 403–404.

54. Khintchine, A., "Korrelationstheorie der stationären stochastischen Prozesse," *Math. Ann.*, Vol. 109, 1934, pp. 604–615.

55. Klein, L., "The Estimation of Distributed Lags," *Econometrica*, Vol. 26, No. 4, October 1958, pp. 553–565.

56. Kolmogorov, A. N., "Sur l'interpolation et l'extrapolation des suites stationaires," *Compt. rend. acad. sci.*, Paris, Vol. 208, 1939, pp. 2043–2045.

57. Kuznets, S. S., *Capital and the American Economy: Its Formation and Financing*, National Bureau of Economic Research, New York, 1961.

58. Lanczos, C., *A Discourse on Fourier Series*, Hafner, New York, 1966.

59. Leviatan, N., "Consistent Estimates of Distributed Lags," *Intern. Econ. Rev.*, Vol. 2, No. 1, January 1963, pp. 44–52.

60. Lomnicki, Z. A., and S. K. Zaremba, "On Estimating the Spectral Density Function of a Stochastic Process," *J. Roy. Stat. Soc.*, Series B, Vol. 19, No. 1, 1957, pp. 13–37.

61. ——, "On Some Moments and Distributions Occurring in the Theory of Linear Stochastic Processes," *Monatshefte für Math.*, Part I, Vol. 61, 1957, pp. 318–358, and Part II, Vol. 63, 1959, pp. 94–118.

62. Malinvaud, E., *Statistical Methods of Econometrics*, Rand McNally, Chicago, 1966.

63. Marriott, F. H. C., and J. A. Pope, "Bias in the Estimation of Autocorrelations," *Biometrika*, Vol. 41, Parts 3 and 4, 1954, pp. 390–402.

64. Moore, H. L., *Economic Cycles: Their Law and Cause*, The Macmillan Company, New York, 1914.

65. Murthy, V. K., "Estimation of the Cross-Spectrum," *Ann. Math. Stat.*, Vol. 34, No. 3, September 1965, pp. 1012–1021.

66. Nerlove, M., *Distributed Lags, and Demand Analysis for Agricultural and Other Commodities*, Agricultural Marketing Service, United States Department of Agriculture, June 1958.

67. ——, "Spectral Analysis of Seasonal Adjustment Procedures," *Econometrica*, Vol. 32, No. 3, July 1964, pp. 241–286.

68. ——, "A Comparison of a Modified 'Hannan' and BLS Seasonal Adjustment Filters," *J. Am. Stat. Assoc.*, Vol. 60, No. 310, June 1965, pp. 442–491.

69. ——, *Distributed Lags and Unobserved Components in Economic Time Series*, Cowles Commission Discussion Paper 211, Yale University, New Haven, Connecticut, March 13, 1967.

70. Nettheim, N. F., *A Spectral Study of "Overadjustment" for Seasonality*, Department of Statistics, Stanford University, Stanford, California, Technical Report 1, prepared under Contract Nonr-225(80), (NR-042-234) for the Office of Naval Research, November 11, 1964.

71. ——, *The Estimation of Coherence*, Department of Statistics, Stanford University, Stanford, California, Technical Report 5, prepared under Contract Nonr-225(80) (NR-042-234) for the Office of Naval Research, May 16, 1966.

72. Parzen, E., "On Asymptotically Efficient Consistent Estimates of the Spectral Density Function of a Stationary Time Series," *J. Roy. Stat. Soc.*, Series B, Vol. 20, No. 2, 1958, pp. 303–322.

73. ——, "Mathematical Considerations in the Estimation of Spectra," *Technometrics*, Vol. 3, No. 2, May 1961, pp. 167–190.

74. ——, "An Approach to Time Series Analysis," *Ann. Math. Stat.*, Vol. 32, No. 4, December 1961, pp. 951–988.

75. ——, "An Approach to Empirical Time Series Analysis," *Radio Sci.*, published by National Bureau of Standards, United States Department of Commerce, Vol. 68D, No. 9, September 1964, pp. 937–952.

76. ——, *The Role of Spectral Analysis in Time Series Analysis*, presented at the 35th Session of the International Statistical Institute, Belgrade, September 1965, pp. 1–25.

77. ———, *Analysis and Synthesis of Linear Models for Time Series*, Department of Statistics, Stanford University, Stanford, California, Technical Report 4, prepared under Contract Nonr-225(80) (NR-042-234) for the Office of Naval Research, April 18, 1966.

78. ———, *Analysis for Models of Signal Plus White Noise*, Department of Statistics, Stanford University, Stanford, California, Technical Report 6, prepared under Contract Nonr-225(80) (NR-042-234) for the Office of Naval Research, September 12, 1966.

79. Rosenblatt, M., "Statistical Analysis of Stochastic Processes," in *Probability and Statistics*, U. Grenander (ed.), John Wiley and Sons, New York, 1959, pp. 246–273.

80. ———, "Some Comments on Narrow Band-Pass Filters," *Quart. Appl. Math.*, Vol. 18, No. 4, January 1961, pp. 387–393.

81. ———, *Random Processes*, Oxford University Press, New York, 1962.

82. Rudnick, P., "Note on the Calculation of Fourier Series," *Math. Comput.*, Vol. 20, 1966, pp. 429–430.

83. Scheffé, H., *The Analysis of Variance*, John Wiley and Sons, New York, 1959.

84. Schmid, C. F., *Handbook of Graphic Representation*, The Ronald Press Company, New York, 1964.

85. Schuster, Sir Arthur, "On the Investigation of Hidden Periodicities With Application to a Supposed Twenty-Six Day Period of Meteorological Phenomena," *Terrestrial Magnetism*, Vol. 3, 1898, p. 13.

86. Shaerf, M. Casani, *Estimation of the Covariance and Autoregressive Structure of a Stationary Time Series*, Technical Report 12, prepared under Grant DA-ARO(D)-31-124-G363 for United States Army Research Office, Department of Statistics, Stanford University, Stanford, California, January 13, 1964 (mimeograph).

87. Singleton, R. C., and T. C. Poulter, "Spectral Analysis of the Call of the Male Killer Whale," *IEEE Transactions on Audio and Electroacoustics*, Vol. AU-15, No. 2, June 1967, pp. 104–113.

88. Slutzky, E., "The Summation of Random Causes as the Source of Cyclic Processes," *Econometrica*, Vol. 5, 1937, pp. 105–146.

89. Tintner, G., *The Variate Difference Method*, Principia Press, Bloomington, Indiana, 1940.

90. Tukey, J. W., "Discussion Emphasizing the Connection Between Analysis of Variance and Spectral Analysis," *Technometrics*, Vol. 3, No. 2, May 1961, pp. 191–220.

91. ———, "An Introduction to the Calculations of Numerical Spectrum Analysis," in *Advanced Seminar on Spectral Analysis of Time Series*, B. Harris (ed.), John Wiley and Sons, New York, 1967.

92. United States Department of Commerce, Office of Business Economics, *Survey of Current Business* (monthly).

93. ———, *Business Statistics, Supplement*, August 1965.

94. Wahba, G., *Estimation of the Coefficients in a Multidimensional Distributed Lag Model*, Department of Statistics, Stanford University, Stan-

ford, California, Technical Report 18, prepared under National Science Foundation Grant GP-4265, January 30, 1967.

95. Walker, A. M., "The Asymptotic Distribution of Serial Correlation Coefficients for Autoregressive Processes with Dependent Residuals," *Proc. Cambridge Phil. Soc.*, Vol. 50, 1954, pp. 60–64.

96. Wallis, K. F., *Distributed Lag Relationships Between Retail Sales and Inventories*, Institute for Mathematical Studies in the Social Sciences, Stanford University, Stanford, California, National Science Foundation Grant, GS-142, Technical Report 14, July 26, 1965.

97. ————, *Some Econometric Problems in the Analysis of Inventory Cycles*, Cowles Commission Discussion Paper 209, Yale University, New Haven, Connecticut, May 9, 1966.

98. Weinstein, A. S., "Alternative Definitions of the Serial Correlation Coefficients in Short Autoregressive Sequences," *J. Am. Stat. Assoc.*, Vol. 53, No. 284, December 1958, pp. 881–892.

99. Welch, B. L., "A Generalization of Student's Problem When Several Different Population Variances are Involved," *Biometrika*, Vol. 34, January 1947, pp. 28–35.

100. Welch, P., "The Use of Fast Fourier Transform for the Estimation of Power Spectra: A Method Based on Time Averaging Over Short, Modified Periodograms," *IEEE Transactions on Audio and Electroacoustics*, Vol. AU-15, No. 2, June 1967, pp. 70–73.

101. Whittle, P., *Prediction and Regulation by Linear Least-Square Methods*, English Universities Press, London, 1963.

102. Wiener, N., "Generalized Harmonic Analysis," *Acta Math.*, Vol. 35, 1930, pp. 117–258.

103. ————, *Extrapolation, Interpolation and Smoothing of Stationary Time Series*, Technology Press, Cambridge, and John Wiley and Sons, New York, 1949.

104. Wilks, S. S., *Mathematical Statistics*, John Wiley and Sons, New York, 1962.

105. Wold, H., *A Study in the Analysis of Stationary Time Series*, Almquist and Wicksells, Uppsala, 1954.

106. Yaglom, A. M., *An Introduction to the Theory of Stationary Random Functions*, trans. from the Russian by R. A. Silverman, Prentice-Hall, Inc., Englewood Cliffs, New Jersey, 1962.

107. Yule, G. U., "On the Time Correlation Problem, with Especial Reference to the Variate-Difference Correlation Methods," *J. Roy. Stat. Soc.*, Vol. 84, 1921, pp. 497–526.

108. ————, "On a Method of Investigating Periodicities in Disturbed Series with Special Reference to Wolfer's Sunspot Numbers," *Phil. Trans. Roy. Soc.*, Vol. 226, Series A, 1927, pp. 267–298.

109. Zaremba, S. K., "Quartic Statistics in Spectral Analysis," in *Advanced Seminar on Spectral Analysis of Time Series*, B. Harris (ed.) John Wiley and Sons, New York, 1967.

Index

Selected RAND Books

Arrow, Kenneth J., and Marvin Hoffenberg, *A Time Series Analysis of Interindustry Demands*, Amsterdam, North-Holland Publishing Company, 1959.

Bellman, Richard E., *Dynamic Programming*, Princeton, N. J., Princeton University Press, 1957.

Bellman, R. E., *Adaptive Control Processes: A Guided Tour*, Princeton, N. J., Princeton University Press, 1961.

Bellman, R. E. (ed.), *Mathematical Optimization Techniques*, Los Angeles, University of California Press, 1963.

Bellman, Richard E., and Kenneth L. Cooke, *Differential-Difference Equations*, New York, Academic Press, Inc., 1963.

Bellman, Richard E., and Stuart E. Dreyfus, *Applied Dynamic Programming*, Princeton, N. J., Princeton University Press, 1962.

Bellman, Richard E., and Robert E. Kalaba, *Quasilinearization and Nonlinear Boundary-Value Problems* (Volume 3), New York, American Elsevier Publishing Company, Inc., 1965.

Dantzig, George B., *Linear Programming and Extensions*, Princeton, N. J., Princeton University Press, 1963.

Dorfman, Robert, Paul A. Samuelson, and Robert M. Solow, *Linear Programming and Economic Analysis*, New York, McGraw-Hill Book Company, 1958.

Dresher, Melvin, *Games of Strategy: Theory and Applications*, Englewood Cliffs, N. J., Prentice-Hall Inc., 1961.

Dreyfus, Stuart, *Dynamic Programming and the Calculus of Variations*, New York, Academic Press Inc., 1965.

Ford, L. R., Jr., and D. R. Fulkerson, *Flows in Networks*, Princeton, N. J., Princeton University Press, 1962.

Gale, David, *The Theory of Linear Economic Models*, New York, McGraw-Hill Book Company, 1960.

Harris, Theodore E., *The Theory of Branching Processes*, Berlin, Springer-Verlag, 1963; New York, Prentice-Hall, 1964.

Hirshleifer, Jack, James C. DeHaven, and Jerome W. Milliman, *Water Supply: Economics, Technology, and Policy*, Chicago, The University of Chicago Press, 1960.

Hitch, Charles J., and Roland McKean, *The Economics of Defense in the Nuclear Age*, Cambridge, Harvard University Press, 1960.

Jorgenson, D. W., J. J. McCall, and R. Radner, *Optimal Replacement Policy*, Chicago, Rand McNally & Company, 1967.

Marschak, T., T. K. Glennan, Jr., and R. Summers, *Strategy for R&D*, New York, Springer-Verlag New York, Inc., 1967.

McKinsey, J. C. C., *Introduction to the Theory of Games*, New York, McGraw-Hill Book Company, 1952.

Meyer, John R., John F. Kain, and Martin Wohl, *The Urban Transportation Problem*, Cambridge, Harvard University Press, 1967.

Novick, David (ed.), *Program Budgeting: Program Analysis and the Federal Budget*, Second Edition, Cambridge, Harvard University Press, 1965.

Quade, E. S. (ed.), *Analysis for Military Decisions*, Chicago, Rand McNally & Company; Amsterdam, North-Holland Publishing Company, 1964.

Quade, E. S., and Wayne I. Boucher (eds.), *Systems Analysis and Policy Planning: Applications in Defense*, New York, American Elsevier Publishing Company, 1968.

Selin, Ivan, *Detection Theory*, Princeton, N. J., Princeton University Press, 1965.

Williams, J. D., *The Compleat Strategyst: Being a Primer on the Theory of Games of Strategy*, New York, McGraw-Hill Book Company, 1954.

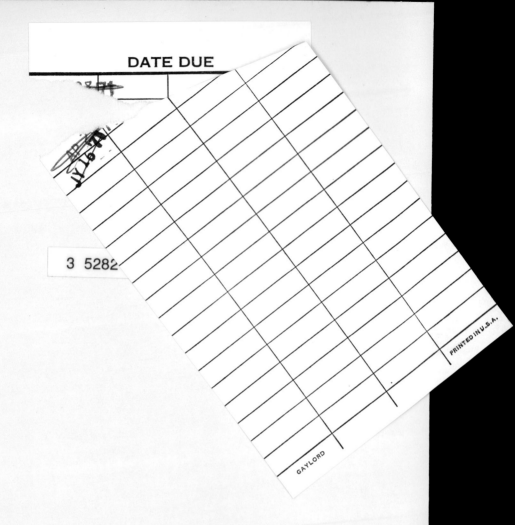

GAYLORD

PRINTED IN U.S.A.